atics for GCSE

Exclusively endorsed and approved by AQA

Revision Guide

Series Editor
Paul Metcalf

June Haighton
Andrew Manning
Kathryn Scott

HIGHER

Linear   Modular

Nelson Thornes

Published in 2007 by:
Nelson Thornes Ltd
Delta Place
27 Bath Road
CHELTENHAM
GL53 7TH
United Kingdom

08 09 10 11 / 10 9 8 7 6 5 4 3 2

A catalogue record for this book is available from the British Library

ISBN 978 0 7487 8193 5

Cover photograph by Oxford Scientific Films
Illustrations by Heelstone Publishing Services and Roger Penwill
Page make-up by Heelstone Publishing Services

Printed and bound in Croatia by Zrinski

Acknowledgement
AQA material is reproduced by permission of the Assessment and
Qualifications Alliance.

Revision Tips

What ...

◎ Use this Revision Guide and your exam specification to see what topics you need to cover. Ask your teacher which specification you are following and download one from the web.

◎ Go through the topics, any past papers and tests and make a list of the areas that you find difficult. Concentrate your revision on these.

◎ Check out which formulae will be given on the exam paper. Your teacher can tell you this. Practise using these and make sure you learn any formulae that aren't given.

When and where ...

◎ Start your revision early. Aim to start at least a month before your exam.

◎ Revise little and often rather than in one long session. After 45 minutes take a 10 minute break.

◎ If your school or college has a revision club then check it out.

◎ Find a quiet, well lit place to study and have all your books handy. Keep some scrap paper nearby to write notes and practise questions.

◎ Avoid distractions such as the television, and don't revise late at night – it's important to be alert.

◎ Eat well, take some exercise and get enough sleep. When you're away from your revision enjoy yourself and don't think about work.

How ...

◎ Create an effective revision timetable.

- Find out the date of your exam.
- Divide your subjects up into topics.
- Mix and match harder and easier topics to break it up a bit.
- Schedule 10 minutes of top up time at the start of each session to look back over the stuff you covered last time.
- Tick off topics as you've done them.
- Timetable in some time off and rewards.
- Allow time in the last week to do exam style questions. Go over everything one last time.
- Stick to your schedule but if you feel comfortable with some topics, shift it around to allow extra time on the harder topics.

◎ Use cards to summarise the key points listed in this book. Condense them to one side of paper and take it everywhere with you, reading it at every opportunity.

◎ Use mnemonics. For example, Never Eat Soggy Waffles helps you remember the points of the compass for bearing questions. Illustrate your key points with diagrams and use highlighter pens.

◎ Write formulae on sticky notes and put them where you'll see them regularly – the fridge, the bathroom mirror, …

◎ Get a revision buddy. Take it in turns to explain topics to each other. Test each other. Discuss what works and doesn't work.

◎ Create mind maps or spider diagrams for different topics using plenty of colour and stick them on your wall.

◎ Don't try to memorise maths; try to understand the processes.

◎ Don't be tempted to spend too long on topics that you find easy; concentrate on things that you find harder or get wrong.

◎ Use the end-of-chapter questions to make sure you've understood everything.

◎ Use the exam style questions to practise, practise, practise.

quantitative

discrete

continuous

sampling

convenience or
opportunity

random

systematic

quota

cluster

stratified

tally chart

two-way table

◎ **Qualitative data** is data that cannot be measured using numbers.
For example, hair colour, favourite film, type of car, etc.

◎ **Quantitative data** can be measured using numbers.
For example, number of DVDs, collar size, height, weight, temperature, age, etc and can be split into two categories: discrete and continuous.

◎ **Discrete data** can only take exact numbers, for example, shoe sizes as you can have $3\frac{1}{2}$ and 4 but nothing in between.

◎ **Continuous data** can take any values, for example, length as you can have 10 cm and 11 cm and any number of measurements in between.

Sampling methods

When undertaking surveys, you will need to decide a sampling method. There are six different types of sampling methods that you should be familiar with:

> Surveys might include direct observation or personal surveys (for example, face-to-face interviews or telephone, postal and internet surveys). You will need to be able to design surveys and questionnaires for the examination.

◎ **Convenience** or **opportunity sampling** means that you just take the first people who come along or those who are convenient to sample (such as friends and family).

◎ **Random sampling** requires each member of the population to be assigned a number and the sample is chosen using random numbers.

◎ **Systematic sampling** involves taking every nth member of the population where n is chosen by dividing the population size by the sample size.

◎ **Quota sampling** involves choosing a sample with certain characteristics. For example, selecting 20 adult men, 20 adult women, 10 teenage girls and 10 teenage boys.

◎ **Cluster sampling** splits the population into smaller groups (clusters) and takes a random selection of clusters and surveys all members within them.

◎ **Stratified sampling** involves dividing the population into a series of groups or *strata* and then taking a random sample from each of these.

Worked Example **Classifying data**

Question

Say whether each of the following are qualitative or quantitative.

Where the answer is quantitative state whether the data is discrete or continuous.

a Temperature in a room

b Number of cars at a junction

c Colours of cars at a junction

d Cost of mobile phones

GET IT RIGHT!

It is easy to confuse qualitative and quantitative – remember that **qualitative** goes with **quality** and **quantitative** goes with **quantity**.

Solution

a Temperature in a room is quantitative and continuous.

b Number of cars at a junction is quantitative and discrete.

c Colours of cars at a junction is qualitative.

d Cost of mobile phones is quantitative and discrete.

GET IT RIGHT!

It is also easy to confuse discrete and continuous – remember, for example, that shoe size is discrete but shoe length is continuous.

Worked Example **Questionnaire design**

Question

Frank is writing a questionnaire about computer games.

This is one question from Frank's questionnaire:

GET IT RIGHT!

Remember that the best questionnaires are:

- **appropriate** to the survey being carried out and do not ask unnecessary questions
- **unbiased** so they do not lead the respondent to give a particular answer
- **unambiguous** so they are clear and straightforward to the respondent.

> How much money do you spend on computer games?
>
> Tick one of the boxes.
>
> £0–10 ☐   £10–20 ☐   £20–30 ☐

Write down two criticisms of Frank's question.

AQA EXAMINER SAYS...

When designing questionnaires keep questions short and simple, include time scales, avoid sensitive issues and do not ask personal questions.

Solution

The question does not give a time scale so it is not clear whether it is per week, per month etc.

The intervals overlap so it is not clear where £10 or £20 would go.

Worked Example **Representing data in tally charts**

Question

The following information shows the numbers of brothers and sisters members of a form group have.

0	5	0	1	1	5
1	2	1	2	2	3
1	1	2	2	1	2
1	0	6	0	1	0

In this form, the information is called <u>raw data</u> because it has still to be organised.

Complete a tally chart and a frequency distribution for this data.

5	||	2
6	|	1

24

Always add up the frequencies to check you have the correct total.

Worked Example Representing data in two-way tables

Question

Another form group records their information as a two-way table.

Brothers and sisters

	0	1	2	3	4	5	6
Girls	3	5	4	2	1	0	0
Boys	2	3	3	0	1	0	1

A two-way table is useful to identify different groups.

The total for the rows and the total for the columns should be the same – you may wish to add these to the table.

a How many boys are there altogether in the class?

b How many girls have four brothers and sisters altogether?

c What fraction of the class has no brothers and sisters?

d What is the maximum number of children in any family in the form group?

Solution

a There are 10 boys altogether

b 1 girl has four brothers and sisters altogether

c Fraction of the class is $\frac{5}{25} = \frac{1}{5}$

d 7

GET IT RIGHT!

Think carefully about the wording of questions such as this. The table says that 1 boy has 6 brothers and sisters altogether so there are 7 children in his family.

Worked Example Representing grouped data

Question

For larger data sets it may be necessary to group data, as in this example.

The following information shows the reaction times of students in seconds, to the nearest second.

8	4	15	21	12	10	19	15	7	18	16	9
21	5	7	13	12	21	13	19	23	11	18	20

Complete a grouped tally chart and a frequency distribution for this data

Solution

When choosing suitable groups it is best to have between four and six groups that are all the same size.

Time (secs)	Tally	Frequency
$0 < h \leqslant 5$	||	2
$5 < h \leqslant 10$	卌	5
$10 < h \leqslant 15$	卌 ||	7
$15 < h \leqslant 20$	卌 |	6
$20 < h \leqslant 25$	||||	4

 **GET IT RIGHT!**

For grouped tally charts ensure that the tallies are put in the correct intervals, for example, 10 goes in $5 < h \leqslant 10$ not $10 < h \leqslant 15$

Worked Example Sampling data

Question

The table shows the number of students in each year group of a 14–18 school.

Year	9	10	11	12	13
Number	210	210	240	110	130

Explain how you would obtain a stratified sample of 50 students for a survey.

Remember to show your working.

A stratified sample is useful to ensure that the sample is representative of the population as a whole.

Solution

To take a stratified sample you need to appreciate that 210 students out of 900 students are from Year 9, so $\frac{210}{900}$ of the sample should be from Year 9.

If the required sample size is 50 then $\frac{210}{900} \times 50$ will be from Year 9, that is, 12 students from Year 9 (rounding up the answer).

Completing this information for each year group:

Year	9	10	11	12	13
Number	210	210	240	110	130
Fraction	$\frac{210}{900}$	$\frac{210}{900}$	$\frac{240}{900}$	$\frac{110}{900}$	$\frac{130}{900}$
For a sample size of 50	$\frac{210}{900} \times 50$ = 12 students	$\frac{210}{900} \times 50$ = 12 students	$\frac{240}{900} \times 50$ = 13 students	$\frac{110}{900} \times 50$ = 6 students	$\frac{130}{900} \times 50$ = 7 students

 **GET IT RIGHT!**

Remember to check that the totals for each year group do add to the sample size (in this case, 50).

 **BUMP UP A/A***

THE GRADE To get a grade **A/A*** you will need to be familiar with different types of sampling techniques, especially stratified sampling.

Collecting data

1 For each of the following say whether the data is quantitative or qualitative.

   a The number of people at a rugby match.

   b How many tins of beans a shop sells.

   c The flavour of the beans.

   d The time it takes to travel from London to Manchester.

2 For each of the following say whether the data is discrete or continuous.

   a The number of votes for a party at a local election.

   b The number of beans in a tin.

   c The weight of a tin of beans.

   d The time taken to complete this chapter.

3 The two-way table shows the colour and make of cars in a car park.

	Red	Blue	Black	White
Ford	3	5	1	2
Vauxhall	2	4	3	0
Toyota	1	2	0	2
Other	2	2	3	3

Use the table to answer the following questions:

   a How many red cars were there in the car park?

   b How many Vauxhall cars were there in the car park?

   c What percentage of the cars were black?

4 A college wishes to undertake a survey on its sports facilities. Explain how you would take:

   a a random sample of 50 students

   b a systematic sample of 50 students.

5 Kazama works in a factory that produces batteries.

   He is asked to test a sample of 100 batteries stratified by voltage.

   The table shows the number of each type of battery produced each day.

Battery	AAA	AA	9V	C	D
Number of batteries	3500	6000	700	400	400

   Explain how Kazama should obtain his stratified sample.

   Remember to show your working.

2 Statistical measures

Key words

discrete data

continuous data

mode

modal class

median

mean

range

grouped data

frequency table

frequency distribution

lower bound

upper bound

Key points

🌀 The **mode** is the value that occurs most often.

🌀 The **median** is the middle value when the values are arranged in **order of size**.

GET IT RIGHT!

When finding the median, you must remember to put the numbers in order of size.

> *If there are two middle values, add the two values together and then divide by two.*

🌀 The **mean** is found by calculating $\dfrac{\text{the total of all the values}}{\text{the number of values}}$

🌀 For a **grouped** distribution, the **mean** is found by calculating $\dfrac{\text{the total of all the (frequencies} \times \text{values)}}{\text{the total of frequencies}}$

> *or the mean $= \dfrac{\Sigma fx}{\Sigma f}$ where Σ means 'the sum of'*

🌀 The **range** is the difference between the largest and smallest numbers.

Worked Example **Frequency distributions**

(Question) The numbers of students in 25 classes are shown in the table below.

Number of students in a class (x)	Frequency (f)
25	1
26	2
27	4
28	4
29	6
30	4
31	2
32	2

Work out the mean, mode, median and range of the data.

Solution

Completing the table:

Number of students in a class (x)	Frequency (f)	Frequency × students (fx)
25	1	25 × 1 = 25
26	2	26 × 2 = 52
27	4	27 × 4 = 108
28	4	28 × 4 = 112
29	6	29 × 6 = 174
30	4	30 × 4 = 120
31	2	31 × 2 = 62
32	2	32 × 2 = 64
Total	$\sum f = 25$	$\sum fx = 717$

> Start by adding an extra column to work out 'fx'.
> Complete the totals to give $\sum f$ and $\sum fx$.

$\sum f = 1 + 2 + 4 + 4 + 6 + 4 + 2 + 2$

$\sum fx = 25 + 52 + 108 + 112 + 174 + 120 + 62 + 64$

EXAMINER SAYS...
The examination paper will have sufficient space for you to add further columns. Don't waste time copying out the table again.

GET IT RIGHT!
In this example, 6 is the highest frequency so the mode is 29 as this is the value that occurs **most** often.

Mean $= \dfrac{\sum fx}{\sum f} = \dfrac{717}{25} = 28.68$

Mode $= 29$

The median is the **middle** value when the values are arranged in order of size.

> If there are n values then <u>the middle value</u> is the $\dfrac{(n + 1)\text{th}}{2}$ value.

Since there are 25 values then the median is the $\dfrac{(25 + 1)\text{th}}{2} = 13$th value.

Number of students in a class (x)	Frequency (f)	Cumulative frequency (Σf)	
25	1	1	=1
26	2	1+2	=3
27	4	1+2+4	=7
28	4	1+2+4+4	=11
29	6	1+2+4+4+6	=17
30	4		
31	2		
32	2		

> If you add up the frequencies you get the cumulative frequency (Σf).
> For example, $\Sigma f = 3$ tells you that, altogether, there are 3 classes with up to 26 students in them.
> $\Sigma f = 11$ tells you there are 11 classes with up to 28 students in them.

GET IT RIGHT!
The range should always be given as a single number. Make sure you subtract the values (32 – 25) and not the frequencies (6 – 1).

Median = 29

Range 32 – 25 = 7

The 13th value is in this group (29 students per class) so 29 is the median.

GET IT RIGHT!
When working with grouped data, remember to consider whether the data is discrete or continuous so you can decide on the lower and upper bounds:
Discrete data – Data that can only take individual values is called discrete data. For example, the number of cars is discrete data, you cannot have 2.3 cars!
Continuous data – Continuous data can have any numerical value. Measurements like length and weight are examples of continuous data.

Worked Example Grouped frequency distributions

(Question)

The heights of 80 students to the nearest cm are given in the table below.

EXAMINER SAYS...

On the examination paper, groups can be written in different ways such as 100 up to 110, $120 < x \leq 130$, 140 – .....

Height in cm	Number of students
150–154	3
155–159	4
160–164	9
165–169	16
170–174	18
175–179	17
180–184	7
185–189	6

a What is the modal class of the distribution?

b What are the lower and upper bounds of the first group?

c What is the estimated mean height of the students?

d Which class interval contains the median?

Start by adding extra columns for the midpoint at the mid-interval value and to work out 'fx'.

Complete the totals to give $\sum f$ and $\sum fx$.

(Solution)

AQA EXAMINER SAYS...

The examination paper will have sufficient space for you to add further columns. Don't waste time copying out the table again.

Completing the table:

Height in cms	Number of students	Midpoint (x)	Midpoint × frequency (fx)
150–154	3	152	152 × 3 = 456
155–159	4	157	157 × 4 = 628
160–164	9	162	162 × 9 = 1458
165–169	16	167	167 × 16 = 2672
170–174	18	172	172 × 18 = 3096
175–179	17	177	177 × 17 = 3009
180–184	7	182	182 × 7 = 1274
185–189	6	187	187 × 6 = 1122
Total	$\sum f = 80$		$\sum fx = 13\,715$

GET IT RIGHT!

When finding the mean of a grouped frequency distribution, remember to use the midpoints of the intervals.
152 is halfway between 150 and 154;
157 is halfway between 155 and 159 etc.

$\sum f = 3 + 4 + 9 + 16 + 18 + 17 + 7 + 6$

$\sum fx = 456 + 628 + 1458 + 2672 + 3096 + 3009 + 1274 + 1122$

You cannot find the mode of a grouped frequency distribution, but the modal class is the interval with the highest frequency.

a The modal class is 170–174 cm.

b The first group is 150–154 cm so the lower bound is 149.5 cm and the upper bound is 154.5 cm.

Remember that the 150–154 group can take any value in the range 149.5 to 154.5 as they a given to the nearest cm.

c Mean $= \dfrac{\sum fx}{\sum f} = \dfrac{13\,715}{80} = 171.4375 = 171$ cm (3 s.f.)

EXAMINER SAYS...

On the examination paper, you should round your answers to an **appropriate degree of accuracy**.

d The median is the middle height when the heights are arranged in order of size.

Since there are 80 values, the median is the $\dfrac{(80 + 1)\text{th}}{2}$ = 40.5th value

> The 40.5th value is exactly half way between the 40th value and the 41st value.

EXAMINER SAYS...

On the examination paper, you will often be asked why the mean is only an estimate. Remember that in using mid-interval values you are approximating so the answer cannot be exact.

Height in cm	Number of students (f)	Midpoint (x)	Cumulative frequency	
150–154	3		3	= 3
155–159	4		3 + 4	= 7
160–164	9		3 + 4 + 9	= 16
165–169	16		3 + 4 + 9 + 16	= 32
170–174	18		3 + 4 + 9 + 16 + 18	= 50
175–179	17			
180–184	7			
185–189	6			

The 170–174 class interval contains the median height.

The **40.5th value** is exactly halfway between the 40th value and the 41st value.

Statistical measures

END OF CHAPTER QUESTIONS

Time Yourself!

Can you complete these questions in **20** minutes?

1 A teacher records the number of homeworks completed by 11 students as follows:

19   10   17   18   19   18   17   22   7   13   18

Calculate the mean, median, mode and range.

2 The mean age of five friends is 36.

Another person joins the group and the mean age drops to 35.

How old is the sixth person?

3 Becky asks 15 people how many people are in their family.

Her results are shown in the table

Number in family	2	3	4	5	6	7	8
Frequency	1	3	6	3	1	0	1

From this information calculate the median, mode, mean and range.

4 A receptionist monitors the waiting times for patients in a surgery.

Time (nearest min.)	3–5	6–8	9–11	12–14	15–17	18–22	23–27	28–40
Frequency		43	51	27	11	5	2	0	1

a What is the modal class of the distribution?

b What are the lower and upper bounds of the modal class?

c What is the estimated mean length of the wait?

d Which class interval contains the median?

Representing data

Key words

pie chart

line graph

stem-and-leaf diagram

time series

moving average

cumulative frequency diagram

box plot or box and whisker plot

histogram

frequency polygon

Key points

You will need to be familiar with the following types of representations:

Pie chart

In a pie chart, the frequency is shown by the angles (or areas) of the sectors of a circle.

Mobile phone sales

Time series

A time series is a line graph showing values over time. A time series can be used to predict trends.

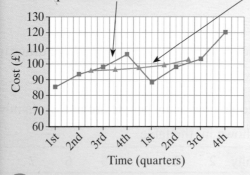

Box plot

A box plot (or box and whisker plot) is used to compare distributions and is always represented in the following format:

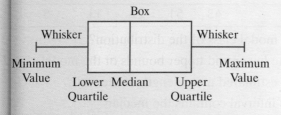

Stem-and-leaf diagram

In a stem-and-leaf diagram, each value is represented as a stem (tens) and leaf (units).

Number of minutes to complete homework

Stem (tens)	Leaf (units)
1	8 9 9
2	3 5 7 8 9
3	3 7 8

In this case, the number 7 stands for 37 (3 tens and 7 units), that is, Key 3|7 represents 37

Moving average

A moving average is used to 'smooth' out the fluctuations in a time series.

A four-point moving average is found by averaging successive four points at a time. They can be plotted on the time series to show the trend.

Cumulative frequency diagram

A cumulative frequency diagram is found by adding the frequencies to give a running total. A cumulative frequency diagram can be used to find an estimate for the median and quartiles of a set of data.

Cumulative frequency diagram

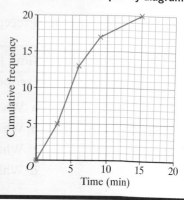

◎ Histogram

A histogram is similar to a bar graph except that the areas of the bars represent the frequencies.

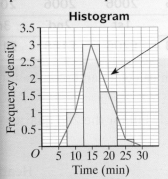

Histogram

◎ Frequency polygon

A frequency polygon can be drawn from a histogram by joining the midpoints of the top of each bar with straight lines.

In a histogram, the areas of the bars represent the frequencies, so

frequency = class width × height

or $\text{height} = \dfrac{\text{frequency}}{\text{class width}}$

Worked Example Stem-and-leaf diagrams

Question

The following information shows the number of minutes (to the nearest minute) taken to complete a puzzle.

Stem (tens)	Leaf (units)
0	8 9 6 7
1	5 3 5 8 5
2	3 3 8 2

In this case, the number 8 stands for 18 (1 ten and 8 units)

In this case, the number 8 stands for 28 (2 tens and 8 units)

Again it is important to provide a key:
Key 1|3 represents 13 minutes

Find:

a the shortest time taken to complete the puzzle

b the median time taken to complete the puzzle.

c the mean time.

EXAMINER SAYS...

Always include a key for your stem-and-leaf diagram. Similarly, if a stem-and-leaf diagram is provided in a question remember to check its key. Other formats of the key include:

85|2 for 852

5|99 for £5.99

4|11 for 4 feet 11 inches

3|55 for 3 hours 55 mins

Solution

For 13 values, the middle value is the 7th value.
In general, for n values the middle value is the $\dfrac{(n+1)\text{th}}{2}$ value.

Here the leaves (units) are arranged numerically.

a The shortest time taken is 06 minutes or 6 minutes.

b To find the median it is important to have the data in numerical order.

This can be done by creating an **ordered stem-and-leaf diagram**.

Stem (tens)	Leaf (units)
0	6 7 8 9
1	3 5 5 5 8
2	2 3 3 8

Key: 1|3 represents 13 minutes

The median time taken to complete the puzzle is 15 minutes.

c Mean time = 202 ÷ 13 = 15.5 min (3 s.f.)

EXAMINER SAYS...

A common mistake when calculating the mean from a stem-and-leaf diagram is to add 8 + 9 + 6 + 7 + 5 + 3 + 5 + ... rather than 8 + 9 + 6 + 7 + 15 + 13 + 15 + ... that is, ignoring the stems of the numbers. Remember to check that your answer is reasonable.

11

Worked Example Time series

(Question)

The following table shows the quarterly heating cost for a house.

Year	2005	2005	2005	2006	2006	2006	2006	2007
Quarter	2nd	3rd	4th	1st	2nd	3rd	4th	1st
Cost	£185	£193	£198	£206	£188	£198	£203	£220

Show the information on a graph, then calculate and plot the four-point moving averages.

Use the trend line to calculate the likely bill for the 2nd quarter of 2007.

(Solution)

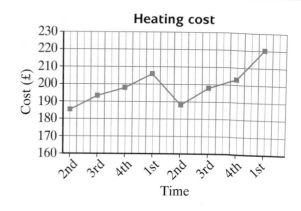

The four-point moving averages can be found by averaging successive four points at a time.

Year	2005	2005	2005	2006	2006	2006	2006	2007
Quarter	2nd	3rd	4th	1st	2nd	3rd	4th	1st
Cost	£185	£193	£198	£206	£188	£198	£203	£220

The **first** four-point moving average $= \dfrac{£185 + £193 + £198 + £206}{4} = £195.50$

The **second** four-point moving average $= \dfrac{£193 + £198 + £206 + £188}{4} = £196.25$

The **third** four-point moving average $= \dfrac{£198 + £206 + £188 + £198}{4} = £197.50$

The **fourth** four-point moving average $= \dfrac{£206 + £188 + £198 + £203}{4} = £198.75$

The **fifth** four-point moving average $= \dfrac{£188 + £198 + £203 + £220}{4} = £202.25$

GET IT RIGHT!

The first four-point moving average is plotted in the 'middle' of the first four points.

The moving average is useful for identifying trends and from this graph you can see that the trend is upwards.

The trend line can be continued to give an indication of future costs.

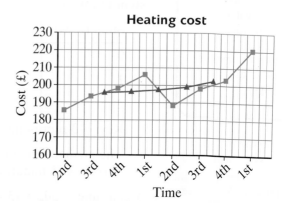

The four-point moving averages can be plotted on the graph as shown.

From the graph, the next four-point moving average would lie around £205.

This assumes that the cost will fall in the 2nd quarter of the following year, based on what happened in the previous years.

The **sixth** four-point moving average $= \dfrac{£198 + £203 + £220 + £x}{4} = £205$

Multiply both sides by 4 $\longrightarrow$ $£198 + £203 + £220 + £x = £205 \times 4$

$£621 + £x = £820$

Subtract £621 from both sides $\longrightarrow$ $£x = £820 - £621$

$£x = £199$

So the next quarterly bill should be about £199.

Worked Example   Cumulative frequency/box plots

Question

The following table shows the times taken to complete a telephone call.

Time (minutes)	Frequency (f)
$0 \leqslant t < 2$	7
$2 \leqslant t < 4$	21
$4 \leqslant t < 8$	16
$8 \leqslant t < 18$	6

The label $2 \leqslant t < 4$ means greater than or equal to 2 and less than 4. The value 4 minutes will not be included in this interval.

a Plot this information as a cumulative frequency diagram.

b Use your graph to calculate an estimate of:

i the median

ii the interquartile range.

c Use this information to draw a box plot of the times taken to complete a telephone call.

Solution

Completing an additional column for the cumulative frequency:

Time (minutes)	Frequency (f)	Cumulative frequency
$0 \leqslant t < 2$	7	7
$2 \leqslant t < 4$	21	28
$4 \leqslant t < 8$	16	44
$8 \leqslant t < 18$	6	50

GET IT RIGHT!

The final cumulative frequency should be the same as the total of the frequencies.

The cumulative frequency is added to the table to provide a 'running total', so 7 + 21 = 28 and 7 + 21 + 16 = 44 etc.

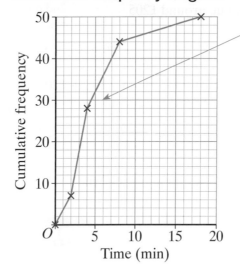

a Cumulative frequency diagram

GET IT RIGHT!

The cumulative frequencies are plotted at the **end** of the intervals.
The cumulative frequency for the $2 \leqslant t < 4$ interval is 28 so is plotted at (4, 28).

Upper quartile is read off at the halfway point in the top half of the data set.

Median is read off at the halfway point in the whole of the data set.

Lower quartile is read off at the halfway point in the bottom half of the data set.

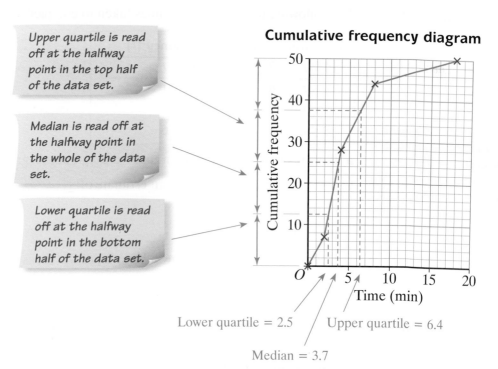

Cumulative frequency diagram

Lower quartile = 2.5   Upper quartile = 6.4

Median = 3.7

b i the median = 3.7

  ii the interquartile range

      = upper quartile – lower quartile

      = 6.4 – 2.5

      = 3.9

The interquartile range = UQ – LQ is a measure of spread concentrating on the middle 50% of the distribution, thus avoiding extreme (rogue) values at the ends of the range.

A box plot (box and whisker plot) provides a visual summary of information and can be used to compare two or more distributions.

c

Time (min)

Worked Example Histograms

(Question) The following table shows the waiting times to get through to a call centre.

Time (minutes)	Frequency (f)
$0 \leqslant t < 2$	0
$2 \leqslant t < 4$	8
$4 \leqslant t < 6$	11
$6 \leqslant t < 10$	6
$10 \leqslant t < 20$	4

 GET IT RIGHT!

In a histogram, the areas of the bars represent the frequencies and
frequency = class width × height
or **height =** $\dfrac{\text{frequency}}{\text{class width}}$

a Plot this information as a histogram.

b Plot this information as a frequency polygon.

c Use your graph to calculate an estimate of the median time.

(Solution) **a** Adding additional columns for class width and frequency density (height of bar):

Time (minutes)	Frequency (f)	Class width	Height = $\dfrac{\text{frequency}}{\text{clas width}}$
$0 \leqslant t < 2$	0	2	Height = $0 \div 2 = 0$
$2 \leqslant t < 4$	8	2	Height = $8 \div 2 = 4$
$4 \leqslant t < 6$	11	2	Height = $11 \div 2 = 5.5$
$6 \leqslant t < 10$	6	4	Height = $6 \div 4 = 1.5$
$10 \leqslant t < 20$	2	10	Height = $2 \div 10 = 0.2$

The height of the bar is called the frequency density.

Frequency density
$= \dfrac{\text{frequency}}{\text{class width}}$

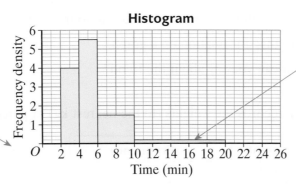

Histogram

Width = 10 and height = 0.2

Frequency = class width × height
$= 10 \times 0.2 = 2$
This bar represents a frequency of 2

A frequency polygon can be drawn from a histogram by joining the midpoints of the top of each bar with straight lines to form a polygon.

b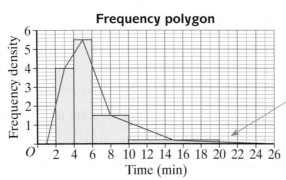

Frequency polygon

The lines should be extended to the horizontal axis so the area under the frequency polygon is equal to the area under the histogram.

15

c

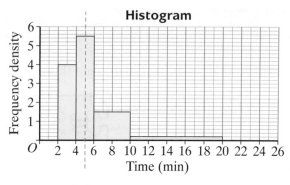

Histogram

Median = 5.0 min

Since the total area under the graph represents the frequency then the median will divide this area into two. The total area = 27 square units so draw the median line so that there are 13.5 square units to the left and 13.5 square units to the right.

THE GRADE
To get a grade **A/A*** you will need to be able to deal confidently with histograms (and frequency polygons).

Time Yourself!

Can you complete these questions in **40** minutes?

Representing data

1 The following table shows the cost of gas bills for each quarter of 2006 and 2007.

Year	2006	2006	2006	2006	2007	2007	2007	2007
Quarter	1st	2nd	3rd	4th	1st	2nd	3rd	4th
Cost	£250	£140	£90	£105	£235	£145	£45	£80

a Use the information to calculate the four-point moving averages.

b Show this information on a graph.

c What can you say about the trend?

2 Hassan records the weights (to the nearest kg) of parcels at a post office.

Weight	Frequency
1–2	21
2–4	14
4–8	7
8–15	3

a Show this information as a cumulative frequency diagram.

b Use your cumulative frequency diagram to estimate:

  i the median   ii the interquartile range.

c Use this information to draw a box plot of the weights of parcels.

3 The box plot below shows the ages of all of the people attending a church.

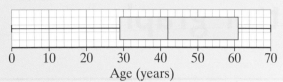

Age (years)

Complete the table to show the information on the box plot.

	Years
Minimum age	
Maximum age	
Lower quartile age	
Upper quartile age	
Median age	

4 The weights of 32 students in a year group are recorded as follows:

Weight (kg)	$40 \leqslant w < 42$	$42 \leqslant w < 45$	$45 \leqslant w < 50$	$50 \leqslant w < 55$	$55 \leqslant w < 65$	$65 \leqslant w < 75$
Frequency	1	3	15	9	3	1

a Show this information as:

i a histogram   **ii** a frequency polygon.

b Use your graph to find the median of the distribution.

4 Scatter graphs

Key words

scatter graph

positive correlation

negative correlation

strong correlation

weak correlation

line of best fit

outlier

Key points

Type of correlation

◎ **Positive correlation** is where an **increase** in one set of data results in an **increase** in the other set of data.

For example, sales of sunglasses against temperature. As the temperature increases, the sales of sunglasses increase.

◎ **Negative correlation** is where an **increase** in one set of data results in a **decrease** in the other set of data.

For example, sales of gloves against temperature. As the temperature increases, the sales of gloves decrease.

◎ **Zero or no correlation** is where there is **no obvious relationship** between the two sets of data.

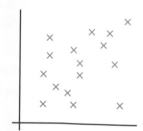

For example, sales of watches against temperature. There is no obvious relationship between temperature and watch sales.

◎ A **line of best fit** is drawn to represent the relationship between two sets of data.

◎ **Strong correlation** is where the points lie close to a straight line. A straight line represents perfect correlation.

◎ **Weak correlation** is where the points lie roughly along a straight line but are not very close to it.

outlier

◎ **No correlation** is where the points are distributed across the whole diagram with no obvious pattern.

◎ A **rogue value**, or **outlier**, is one that does not fit the general trend.

Worked Example Scatter graphs

(Question) The table shows the average temperature and the sales of swimwear for six months in one year.

Month	Jan	Mar	May	Jul	Sep	Nov
Average temperature	6°C	9°C	12°C	18°C	15°C	10°C
Sales of swimwear	2	12	25	47	39	15

Draw a scatter graph to show this information.

Draw a line of best fit on your scatter graph.

Describe the relationship shown by your scatter graph

(Solution)

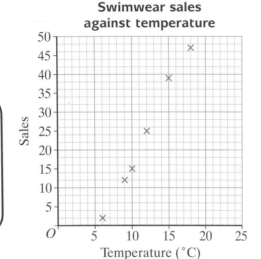

Swimwear sales against temperature

GET IT RIGHT!

Plot your points carefully and check the scales to ensure that they are all correct.

Adding a line of best fit:

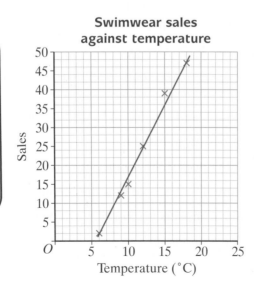

Swimwear sales against temperature

GET IT RIGHT!

You should draw the line of best fit to show the general trend so that roughly equal numbers of points are above and below the line.

The scatter graph shows strong positive correlation between the temperature and sales of swimwear.

A scatter graph is a graph used to show the relationship between two sets of data. You can see from the graph that as the temperature increases, the sales of swimwear increase.

A line of best fit is drawn to represent the relationship between two sets of data. A line of best fit should only be drawn where the correlation is strong.

For additional accuracy, the line of best fit should pass through the point $(\bar{x}, \bar{y})$ where $\bar{x}$ is the mean of all the x-values and $\bar{y}$ is the mean of all the y-values.

AQA

EXAMINER SAYS...

It is important that when you describe the correlation you remember to comment on the **type of correlation** and the **strength of correlation**. You will not gain full marks unless you make reference to both of these in your answer.

Scatter graphs

Time Yourself!

Can you complete these questions in **35** minutes?

1 Describe the relationship shown in each of the following scatter graphs.

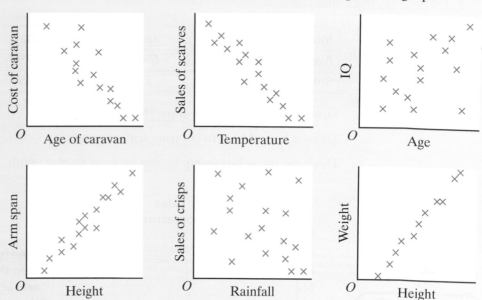

2 The table shows the age and second-hand value of cars.

Age of car (years)	3	1	4	10	6	9	8
Value of car (£)	3300	4800	2300	750	2600	300	1600

a Draw a scatter graph of the results.

b Describe the relationship shown by your scatter graph.

3 Annette collected the following information on the temperature and the number of visitors to a theme park

Temperature (°C)	24	22	20	16	19	23	15	18	21	26
Number of visitors	720	480	440	240	510	550	280	500	600	700

a Draw a scatter graph and the line of best fit of the results.

b Use your line of best fit to estimate

   i the number of visitors if the temperature is 17°C

   ii the temperature if 150 people visit the theme park.

c Which of the above answers is most likely to be inaccurate?

Give a reason for your answer.

5 Probability

Key words

- probability
- certain
- likely
- unlikely
- impossible
- outcome
- theoretical probability
- experimental probability
- mutually exclusive events
- independent events
- dependent events
- sample space diagram
- tree diagram

Key points

◎ Probabilities can range from **impossible** (probability = 0) to **certain** (probability = 1).

Probabilities can be shown on a number line like this:

```
0                    0.5                    1
|--------|------------|------------|--------|
   impossible  unlikely  evens  likely  certain
```

◎ Probabilities can be **experimental** or **theoretical**.

Experimental probability is the probability arising from some experiment (sometimes called the relative frequency).

Theoretical probability is based upon equally likely outcomes and gives an indication of what should happen in theory.

◎ Probabilities are usually expressed as fractions and the probability of an event is

$$\frac{\text{number of required outcomes}}{\text{total number of possible outcomes}}$$

◎ Events are **mutually exclusive** when they cannot happen at the same time. The sum of the probabilities of all the mutually exclusive events is 1. If event A and event B are mutually exclusive, then P(A or B) = P(A) + P(B).

◎ Events are **dependent** when the outcome of one affects the outcome of the other.

◎ Events are **independent** when the outcome of one does not affect the outcome of the other. If event A and event B are independent, then P(A and B) = P(A) × P(B).

◎ A table that shows all possible outcomes is called a **sample space diagram**.

◎ A **tree diagram** is a useful tool for calculating probabilities. The probabilities are written on the branches of the tree. The probabilities on each set of branches should add up to 1.

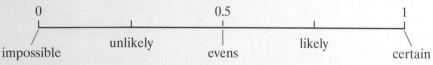

$$\frac{3}{5} \quad G \quad \frac{2}{4} \quad G \quad P(G, G) = \frac{3}{5} \times \frac{2}{4} = \frac{6}{20}$$

$$\frac{2}{4} \quad B \quad P(G, B) = \frac{3}{5} \times \frac{2}{4} = \frac{6}{20}$$

$$\frac{2}{5} \quad B \quad \frac{3}{4} \quad G \quad P(B, G) = \frac{2}{5} \times \frac{3}{4} = \frac{6}{20}$$

$$\frac{1}{4} \quad B \quad P(B, B) = \frac{2}{5} \times \frac{1}{4} = \frac{2}{20}$$

Worked Example **Relative frequency**

Question

A bag contains 5 blue discs, 3 red discs and 2 white discs.
A disc is taken from the bag and the colour noted.
It is then replaced.

a If the experiment is repeated 50 times, how many times would you expect a blue disc to be drawn?

The table shows the results obtained when the experiment is carried out 100 times.

Results from 100 draws			
	Blue	**Red**	**White**
Frequency	54	27	19

The term <u>relative frequency</u> is used to describe experimental probability.

b i What is the relative frequency of a red disc?

ii How does this compare with the theoretical probability?

> **AQA**
> **EXAMINER SAYS...**
> Make sure you understand the difference between relative frequency (from experimental results) and theoretical probability (what should happen in theory). Questions on relative frequency are generally answered badly on the examination.

Solution

a P(blue disc) = $\frac{5}{10}$ Expectation = $\frac{5}{10} \times 50 = 25$

b i Relative frequency of a red disc = $\frac{27}{100}$

The expected frequency is found by multiplying the probability by the number of trials.

ii Expected probability of a red disc = $\frac{3}{10} \times 100 = 30$

The relative frequency is slightly lower.

In this case, the relative frequency is $\frac{27}{100}$ whereas the theoretical probability = $\frac{3}{10} = \frac{30}{100}$

Worked Example **Mutually exclusive events**

Question

A card is drawn from a pack of 52 cards.
What is the probability that the chosen card is:

a a five

b a red card

c a Queen or a King

d a red card or a Jack?

> *P(Queen or King) = P(Queen) + P(King) because the two events are mutually exclusive.*

> 👍 **GET IT RIGHT!**
> Since the sum of the probabilities of all the mutually exclusive events is 1 then if the probability of something happening is p, the probability of it **not** happening is $1 - p$.

Solution

a P(five) = $\frac{4}{52} = \frac{1}{13}$

b P(red card) = $\frac{26}{52} = \frac{1}{2}$

c P(Queen or King) = P(Queen) + P(King) = $\frac{4}{52} + \frac{4}{52} = \frac{8}{52} = \frac{2}{13}$

d P(red or Jack) = $\frac{28}{52}$

> *P(red or Jack) $\neq$ P(red) + P(Jack) because the two events are NOT mutually exclusive. In this case, the Jack of hearts and the Jack of diamonds would be counted twice.*

A♥	2♥	3♥	4♥	5♥	6♥	7♥	8♥	9♥	10♥	J♥	Q♥	K♥
A♦	2♦	3♦	4♦	5♦	6♦	7♦	8♦	9♦	10♦	J♦	Q♦	K♦
										J♣		
										J♠		

Worked Example Independent events

A bag contains 4 red counters and 5 white counters.
One counter is taken at random from the bag.
A second counter is then taken at random from the bag.

a What is the probability of drawing two red counters?

b What is the probability of drawing a red and a white counter?

Solution

An exam question will always say whether the first counter is replaced before the second counter is taken out, so let's look at both cases.

> The two events are
> <u>independent</u> because
> the colour of the first
> counter does not affect
> what colour the second
> counter is.

1 With replacement

Showing this information on a tree diagram:

1st counter   2nd counter

R $\frac{4}{9}$ — R P(R, R) = $\frac{4}{9} \times \frac{4}{9} = \frac{16}{81}$

$\frac{4}{9}$ R

$\frac{5}{9}$ — W P(R, W) = $\frac{4}{9} \times \frac{5}{9} = \frac{20}{81}$

$\frac{4}{9}$ — R P(W, R) = $\frac{5}{9} \times \frac{4}{9} = \frac{20}{81}$

$\frac{5}{9}$ W

$\frac{5}{9}$ — W P(W, W) = $\frac{5}{9} \times \frac{5}{9} = \frac{25}{81}$

> The first counter was
> replaced so it did not
> affect the outcome (or
> the probabilities) when the
> second counter was taken
> out. The two branches
> from each stem always
> add up to one, because
> they show all the possible
> outcomes.

a P(R, R) = $\frac{4}{9} \times \frac{4}{9} = \frac{16}{81}$

b P (red and white) = P(R,W) + P(W, R) = $\frac{20}{81} + \frac{20}{81} = \frac{40}{81}$

You can get a red then a white or a white then a red.

Note that these 4
probabilities add to
give 1

> The two events are
> <u>dependent</u> because
> the colour of the first
> counter does affect
> what colour the
> second counter is.

2 Without replacement

Showing this information on a tree diagram:

1st counter   2nd counter

$\frac{3}{8}$ — R P(R, R) = $\frac{4}{9} \times \frac{3}{8} = \frac{12}{72}$

$\frac{4}{9}$ R

$\frac{5}{8}$ — W P(R, W) = $\frac{4}{9} \times \frac{5}{8} = \frac{20}{72}$

$\frac{4}{8}$ — R P(W, R) = $\frac{5}{9} \times \frac{4}{8} = \frac{20}{72}$

$\frac{5}{9}$ W

$\frac{4}{8}$ — W P(W, W) = $\frac{5}{9} \times \frac{4}{8} = \frac{20}{72}$

You can get a red then a
white or a white then a red.

> The first counter was not
> replaced so it did affect the
> outcome (or the probabilities)
> when the second counter was
> taken out. If the first counter
> is red, there are only 3 red
> counters left and 8 counters
> altogether. If the first counter
> is white, there are only 4 white
> counters left and 8 counters
> altogether.

a P(R, R) = $\frac{4}{9} \times \frac{3}{8} = \frac{12}{72} = \frac{1}{6}$

b P (red and white) = P(R, W) + P(W, R)
= $\frac{20}{72} + \frac{20}{72} = \frac{40}{72} = \frac{5}{9}$

BUMP UP

A/A*

THE GRADE To get a grade **A/A*** you should practise
identifying dependent and independent events and drawing their
tree diagrams to find probabilities.

Probability

Time Yourself!

Can you complete these questions in **30** minutes?

1 The table shows the frequency distribution when a card is drawn from a pack 60 times.

Results from 60 draws			
club	heart	diamond	spade
Frequency distribution 11	17	10	22

   a What is the relative frequency of getting a club?
   b What is the relative frequency of getting a red card?
   c What is the theoretical probability of getting a spade?

2 The probability that Natasha will win a swimming competition is 0.85
   What is the probability that she will not win?

3 A box contains 3 red cubes and 5 blue cubes.
   A cube is taken from the box and then replaced.
   A second cube is then taken from the box.

   Draw a tree diagram and use it to find:

   a the probability that both cubes are red
   b the probability that both cubes are blue
   c the probability that one cube is red and one is blue.

4 A box contains 3 red cubes and 5 blue cubes.
   A cube is taken from the box and not replaced.
   A second cube is then taken from the box.

   Draw a tree diagram and use it to find:

   a the probability that both cubes are red
   b the probability that both cubes are blue
   c the probability that one cube is red and one is blue.

5 The probability that Gail will oversleep is 0.2
   If she oversleeps, the probability that she will be late for school is 0.7
   If she does not oversleep, the probability that she will be late is 0.1

   Draw a tree diagram and use it to find:

   a the probability that Gail does not oversleep and is on time for school
   b the probability that she is late for school.

6 One card is drawn at random from a full pack of 52 cards. It is then replaced and a second card is drawn.

   Draw a tree diagram and use it to find:

   a the probability that both cards are diamonds
   b the probability that neither card is a diamond
   c the probability that just one card is a diamond.

1 Corri works in an electrical shop.
   She is asked to test a sample of 50 light bulbs stratified by type of light bulb.
   The table shows the number of each type of bulb in the shop.

Type of bulb	40 W	60 W	100 W
Number of bulbs	240	680	150

Calculate the number of 60 W bulbs required for her stratified sample.   *(3 marks)*

AQA Spec B, Higher Module 1, Nov 06

2 The table shows the time, in minutes, that 80 customers spent in the queue at a bank.

Time t (minutes)	Frequency
$0 < t \leqslant 1$	2
$1 < t \leqslant 2$	11
$2 < t \leqslant 3$	19
$3 < t \leqslant 4$	31
$4 < t \leqslant 5$	12
$5 < t \leqslant 6$	5

Time t (minutes)	Cumulative frequency
$\leqslant 1$	2
$\leqslant 2$	13
$\leqslant 3$	
$\leqslant 4$	
$\leqslant 5$	
$\leqslant 6$	

(a) Complete the cumulative frequency table.   *(1 mark)*

(b) Draw a cumulative frequency diagram for the data.   *(3 marks)*

(c) Use your graph to estimate the number of customers who spent less
     than 3.5 minutes in the queue.   *(1 mark)*

AQA Spec B, Higher Module 1, June 06

3 The histogram shows the heights of a number of people who wanted to be dancers in a show.

There were 17 people between the heights of 165 cm and 170 cm.
Those people above 185 cm were not accepted for the show.

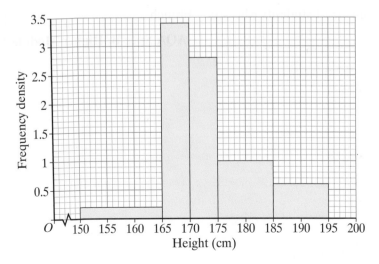

Calculate an estimate of the number of people who were **not** accepted.   *(3 marks)*

AQA Spec B, Higher Module 1, June 06

4 The quarterly heating costs for a house are shown in the table.

Date	Jun 05	Sept 05	Dec 05	Mar 06	Jun 06	Sept 06	Dec 06
Heating costs (£)	50	36	75	79	70	48	

The heating costs and the four-point moving averages are plotted below.

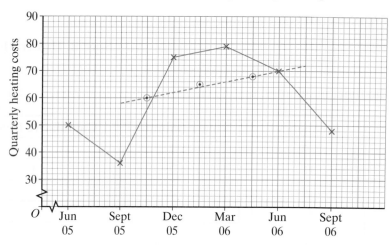

Use the trend line given on the diagram to calculate an estimate of the heating costs for December 2006.

(3 marks)

AQA Spec B, Higher Module 1, Nov 06

5 There are 475 girls in a school.

The probability that a girl studies French is 0.6

(a) How many girls study French?

(2 marks)

(b) Altogether there are 900 students in the school.

Every student studies either French or German.
There are no students who study both.
The probability that a boy studies German is 0.52

Calculate the total number of students who study French.

(3 marks)

AQA Spec B, Higher Module 1, June 06

1 Sophie counts the number of letters in each word of the first sentence of a newspaper.

These are her results.

9	2	3	6	5	7	6
3	7	9	8	4	8	7

There are 14 numbers so the median is the $\frac{(14+1)th}{2}$ value = 7.5th value so between the 7th and 8th value.

(a) Work out the median.

The first mark is awarded for arranging the numbers in order of size.

2   3   3   4   5   6   **6   7**   7   7   8   8   9   9

The candidate identifies that the middle number lies between 6 and 7

Answer **Median = 6.5**   *(2 marks)*

The second mark is awarded here for writing down the correct value for the median as 6.5 or $6\frac{1}{2}$

(b) Calculate the mean of this data.

$$\text{Mean} = \frac{9+2+3+6+5+7+6+3+7+9+8+4+8+7}{14}$$

The first mark is awarded for the correct method of adding the numbers.

$$= \frac{84}{14} = 6$$

The second mark is for dividing by 14 as there are 14 numbers altogether.

It is a good idea to use the original data for adding up – just in case there is a mistake in reordering.

Answer **Mean = 6**   *(3 marks)*

The final mark is awarded for the correct answer of 6

2 The number of minutes that trains arrived late at a station is shown in the table below.

1 mark for correctly finding the midpoints for the intervals

Number of minutes late, t	Frequency	Midpoint	ft
$0 < t \leqslant 10$	16	5	80
$10 < t \leqslant 20$	10	15	150
$20 < t \leqslant 30$	11	25	275
$30 < t \leqslant 40$	8	35	280
$40 < t \leqslant 50$	5	45	225

$$\sum f = 50 \qquad \sum ft = 1010$$

It is useful to add extra columns for the 'ft' values and complete the totals to give $\sum f$ and $\sum ft$

(a) Complete the midpoint column and use it to calculate an estimate of the mean number of minutes that trains arrived late.

The method attracts a second mark making use of $\sum ft = 1010$ and $\sum f = 50$

$$\text{Mean} = \frac{\sum ft}{\sum f} = \frac{1010}{50} = 20.2$$

Answer **Mean = 20.2**   *(3 marks)*

The final mark is awarded for the correct answer of 20.2 – it is best to round this as the answer is only an estimate

(b) Which class interval contains the median number of minutes that trains arrived late?

Number of minutes late, t	Frequency	
$0 < t \leqslant 10$	16	16 = 16
$10 < t \leqslant 20$	10	16 + 10 = 26
$20 < t \leqslant 30$	11	
$30 < t \leqslant 40$	8	
$40 < t \leqslant 50$	5	

The median is the middle value when the values are arranged in order of size. Since there are 50 values then the median is the $\frac{(50+1)\text{th}}{2}$ = 25.5th value.

The median is in the
$10 < t \leqslant 20$ interval.

The 25.5th value is in this group, which contains the 25th and 26th values.

Answer ..... $10 < t \leqslant 20$ ..... minutes

The two marks are awarded for identifying the middle, that is, the 25.5th value, and obtaining the interval from the table.

(2 marks)

3   Phil wants to test if a six-sided dice is biased.
He rolls the dice 20 times.
Here are his results.

2   3   5   6   1   2   4   5   6   2

3   4   2   1   2   3   5   6   2   1

(a) Complete the relative frequency table.

	1	2	3	4	5	6
Tally	III	HHI I	III	II	III	III
Frequency	3	6	3	2	3	3

The candidate appreciates the difference between frequency and relative frequency.

It is helpful to cancel fractions where possible, but to compare fractions it is best if they have the same denominator.

Number	1	2	3	4	5	6
Relative frequency	$\frac{3}{20}$	$\frac{6}{20} = \frac{3}{10}$	$\frac{3}{20}$	$\frac{2}{20} = \frac{1}{10}$	$\frac{3}{20}$	$\frac{3}{20}$

(2 marks)

No marks are given for the tally chart – a method mark is awarded for calculating a relative frequency correctly.

(b) Phil concludes that the dice is biased towards a number.

Write down the number that you think the dice is biased towards.

Explain your answer.

The answer is correct, but the explanation is also required in order to get the mark.

Number ..... 2 .....

Explanation ..... The number 2 occurs much more frequently than the other numbers – the frequencies should all be roughly the same,

(1 mark)

AQA Spec B, Higher Module 1, Nov 06

1

Integers and rounding

Key words

integer
factor
common factor
highest common factor (HCF)
multiple
least common multiple (LCM)
prime number
estimate
approximation
index notation
reciprocal
significant figures
decimal places
upper bound
lower bound

Key points

◎ An **integer** is any positive or negative **whole** number, or zero.

◎ A **factor** of a number is a natural number that divides exactly into it (no remainder). The **highest common factor (HCF)** of two (or more) numbers is the highest number that divides exactly into both (all) of them.

◎ The **multiples** of a number are the products of its multiplication table. The **least common multiple (LCM)** is the lowest multiple that is common to two or more numbers.

◎ A **prime number** is a number with exactly two factors. (Remember that 1 is not a prime number and 2 is a prime number.)

◎ **Index notation** – when $2 \times 2 \times 2 \times 2$ is written as 2^4, the number 4 is the index (plural **indices**).

◎ 1 divided by a number gives the number's reciprocal. For example, the reciprocal of 5 is $\frac{1}{5}$

Any number multiplied by its **reciprocal** equals 1.

◎ The **upper bound** is the maximum possible value of a measurement.

The **lower bound** is the minimum possible value of a measurement.

Worked Example Highest common factor and least common multiple

Question

a Find the highest common factor (HCF) of 18 and 24.

b Find the least common multiple (LCM) of 18 and 24.

Solution

A **factor** of
a number divides
into it.
A **multiple** of
a number is
something it
divides into.

a The factors of 18 are   1   2   3   6   9   18   6 is the **highest** number that is **common to both lists**

The factors of 24 are   1   2   3   4   6   8   12   24

The HCF of 18 and 24 is 6.

b The multiples of 18 are   18   36   54   72   90...   72 is the **lowest** number that is **common to both lists**

The multiples of 24 are   24   48   72   96...

The LCM of 18 and 24 is 72.

Worked Example — Writing a number as a product of its prime factors

Question

a Write 360 as a product of its prime factors in index form.

b Write 3600 as a product of its prime factors in index form.

c When written as the product of prime factors $1120 = 2^5 \times 5 \times 7$

Solution Work out the highest common factor of 1120 and 3600.

GET IT RIGHT!

Prime numbers have exactly two factors (for example, 2, 3, 5, 7, 11, ...)

AQA EXAMINER SAYS...

Make sure you know the meaning of special words: **product** means the result of multiplying numbers.

a

2	360
2	180
2	90
3	45
3	15
5	5
	1

Try the prime numbers in order to see if they are factors.

360 written as a product of its prime factors is $2 \times 2 \times 2 \times 3 \times 3 \times 5$

In index form $360 = 2^3 \times 3^2 \times 5$

The index in 2^3 tells you that 3 twos are multiplied together

b $3600 = 360 \times 10 = 360 \times 2 \times 5$

So $3600 = 2^4 \times 3^2 \times 5^2$

The extra factors 2 and 5, increase the indices of 2 and 5 by 1

c $1120 = 2^5 \times 5 \times 7$ and $3600 = 2^4 \times 3^2 \times 5^2$

HCF of 1120 and 3600 = $2^4 \times 5 = 16 \times 5 = 80$

The HCF is made from the lowest power of each common prime factor, that is 2^4 and 5^1.

The LCM would include the highest power of each prime factor.

LCM of 1120 and 3600 is $2^5 \times 3^2 \times 5^2 \times 7$

Worked Example — Estimating

Question Use approximations to estimate the value of: **a** $\dfrac{49.1 \times 3.06}{0.287}$ **b** $\dfrac{972 + 215}{8.09 - 3.78}$

Solution

Round all numbers to 1 significant figure before doing the calculation.

a $\dfrac{49.1 \times 3.06}{0.287} \approx \dfrac{50 \times 3}{0.3} = \dfrac{150}{0.3} = \dfrac{1500}{3} = 500$

$\times 10$

$\times 10$

b $\dfrac{972 + 215}{8.09 - 3.78} \approx \dfrac{1000 + 200}{8 - 4} = \dfrac{1200}{4} = 300$

AQA EXAMINER SAYS...

You might be asked to do a calculation like part **b** accurately. If so, take care – you need to do the numerator and denominator separately first *or* use brackets, that is, $(972 + 215) \div (8.09 - 3.78)$ gives the correct answer.

AQA EXAMINER SAYS...

Candidates sometimes get no marks because they do an accurate calculation instead of an estimate.

Worked Example Odds, evens and primes

Question

E is an even number. P is a prime number.
State whether each of the following is *always odd* or *always even* or *could be either odd or even*.

a E(P – 1)   **b** P(E – 1)   **c** (E – P)

Solution

a E(P – 1) is *always even*.

b P could be either odd or even (when P = 2) and E – 1 is odd, so P(E – 1) *could be either odd or even*.

c (E – P) *could be either odd or even*.

Worked Example Maximum and minimum values

Question

A plan shows that the length of a rectangular field is 42 metres and its width is 36 metres. Each of these measurements is correct to the nearest metre

a What is the maximum possible perimeter of the field?

b What is the maximum possible difference between the length and the width?

42 m

field   36 m

Solution

For the length, upper bound = 42.5 m and lower bound = 41.5 m

For the width, upper bound = 36.5 m and lower bound = 35.5 m

EXAMINER SAYS...

Take care with upper bounds. For example, if the number of students in a school is 700 to the nearest 10, the upper bound is 704 (whole numbers); if the weight of a cabbage is 700 g to the nearest 10 g, the upper bound is 705 g (measurements).

a The maximum possible perimeter of the field = (42.5 + 36.5) × 2

Use 42.5, not 42.4999... and 36.5, not 36.4999...

= 79 × 2 = 158 metres

Alternatively:
42.5 + 36.5 + 42.5 + 36.5 = 158

b The maximum possible difference between the length and width

= 42.5 – 35.5 = 7 metres

Think carefully about which bounds to use in calculations.
The **difference** will be greatest when the length is as **large** as possible, but the width is as **small** as possible.

THE GRADE   To get a grade **A/A*** you must be able to combine upper and lower bounds in calculations involving +, –, × and ÷

Integers and rounding

Time Yourself!

Can you complete these questions in **50** minutes?

1 Find all the common factors of **a** 24 and 32 **b** 36 and 48

2 Find the highest common factor of
a 20 and 36 **b** 30 and 45 **c** 48 and 72.

3 Find the least common multiple of
a 6 and 15 **b** 12 and 20 **c** 20 and 36

4 Write each number as the product of its prime factors in index notation.
a 28 **b** 54 **c** 90 **d** 120 **e** 250 **f** 567

5 Find the reciprocal of **a** 8 **b** $\frac{1}{3}$ **c** 0.4 **d** $\frac{5}{8}$ **e** −4

6 a Write each number as the product of its prime factors in index notation.
i 72 **ii** 270
b Find the highest common factor (HCF) of 72 and 270.
c Find the least common multiple (LCM) of 72 and 270.

7 Use approximations to estimate the value of the following calculations.
You **must** show your working.

a $\dfrac{206 \times 4.91}{1.92^2}$ **b** $\dfrac{5.74 + 3.12}{0.879}$ **c** $\dfrac{83.9}{19.3 \times 0.163}$ **d** $\dfrac{86.71 - 32.04}{41.56 + 79.52}$

8 The weights of two suitcases are 15 kilograms and 20 kilograms, each correct to the nearest kilogram.

a Find the maximum possible total weight.

b What is the maximum possible difference between the weights?

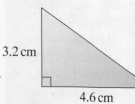

15 kg 20 kg

9 The dimensions of this triangle are correct to 1 decimal place.

Find upper and lower bounds for the area of the triangle.

3.2 cm

4.6 cm

10 The capacity of a paddling pool is 800 litres, correct to the nearest 10 litres.

A pump is used to fill the pool with water at a rate of 30 litres per minute, to 2 significant figures.

Find the maximum and minimum possible times that it takes to fill the paddling pool.

Make sure you can also do all the decimal and fraction calculations on a calculator.

Key points

- To **write a decimal as a fraction**, use the last place value, for example, $0.379 = \frac{379}{1000}$

  To **write a fraction as a decimal**, divide the numerator by the denominator.

- A **recurring** decimal has a repeating digit or group of digits, for example, $0.7305305305\ldots$ written as $0.7\dot{3}0\dot{5}$

- To **add** or **subtract** decimals, **line up the decimal points**.

- When a decimal is **multiplied or divided by 10**, the digits move **1 place**, when it is **multiplied or divided by 100**, the digits move **2 places** etc.

- To **multiply decimals**, remove the decimal points, multiply as usual, then count the number of decimal places.

- To **divide by a decimal**, write as a fraction, multiply numerator and denominator by 10 until the denominator becomes a whole number, then divide as usual.

- To **simplify a fraction**, divide the numerator and denominator by the same numbers until you find the simplest equivalent form.

- To **put fractions in order** or to **add** or **subtract fractions**, write them with the same denominator.

- To **find a fraction of something**, divide it by the denominator, then multiply by the numerator.

- To **multiply fractions**, multiply the numerators and multiply the denominators. (Change mixed numbers to improper fractions first.)

- To **divide by a fraction**, multiply by its reciprocal (upside-down) fraction.

  The **reciprocal** of $\frac{3}{2}$ is $\frac{2}{3}$; the reciprocal of 0.2 is $\frac{1}{0.2} = \frac{10}{2} = 5$.

  $$\overset{\times 10}{\frown}$$
  $$\underset{\times 10}{\smile}$$

- A **rational** number is a number that can be written in the form $\frac{p}{q}$ where p and q are integers, for example, $2, -3, \frac{4}{7}, 3\frac{1}{4}, 4.2, 0.\dot{6}$ are all rational. When rational numbers are written as decimals, they terminate or recur.

- An **irrational** number cannot be written as a fraction.

  For example $\pi, \sqrt{2}, \sqrt[3]{5}$ are all irrational. They are equivalent to non-terminating decimals and do not recur.

- A **surd** is a number that contains an irrational root.

  For example, $\sqrt{2}$ or $5 - \sqrt{2}$.

Worked Example Decimal calculations

Question   Work out: **a** $0.3 \times (1.7 + 0.84)$ **b** $37 - 2.52 \div 0.6$

Solution

GET IT RIGHT!

Remember the order of operations:
Brackets
Indices
Divide
Multiply
Add
Subtract

a
$$\begin{array}{r} 1.70 \\ + 0.84 \\ \hline 2._1 54 \end{array}$$

To add, line up the decimal points.

So $0.3 \times (1.7 + 0.84) = 0.3 \times 2.54$

$$\begin{array}{r} 254 \\ \times \quad 3 \\ \hline 7_16_2 \end{array}$$

To multiply, ignore the decimal points. After multiplying, count the number of decimal places (3 here) so the answer has 3 d.p.

So $0.3 \times (1.7 + 0.84) = 0.762$

b $2.52 \div 0.6 = \dfrac{2.52}{0.6} = \dfrac{25.2}{6}$
(×10)

÷ before −

$$6\overline{)25.^12}\quad = 4.2$$

So $37 - 2.52 \div 0.6 = 37 - 4.2$

$$\begin{array}{r} 3^67.^10 \\ - \quad 4.2 \\ \hline 32.8 \end{array}$$

To subtract, line up the decimal points. Put a 0 in the space.

So $37 - 2.52 \div 0.6 = 32.8$

Worked Example Comparing fractions

Question   Which of the following fractions is nearest to $\frac{2}{3}$? Show how you decide.

$$\frac{5}{6} \qquad \frac{7}{12} \qquad \frac{5}{8} \qquad \frac{3}{4}$$

Change each fraction to an equivalent fraction with denominator 24

Solution   The denominator of the fraction, $\frac{2}{3}$, and the other fractions are all factors of 24.

$$\frac{2}{3} = \frac{16}{24} \quad (\times 8) \qquad \frac{5}{6} = \frac{20}{24} \quad (\times 4) \qquad \frac{7}{12} = \frac{14}{24} \quad (\times 2) \qquad \frac{5}{8} = \frac{15}{24} \quad (\times 3) \qquad \frac{3}{4} = \frac{18}{24} \quad (\times 6)$$

You also use this method to put fractions in order of size.

$\frac{15}{24}$ is nearest to $\frac{16}{24}$, so $\frac{5}{8}$ is the nearest to $\frac{2}{3}$

Alternatively, the fractions could be compared by writing them as decimals.

Worked Example Writing a fraction as a decimal

Question   Write $\frac{8}{11}$ as a decimal.

To write a fraction as a decimal, divide the top by the bottom.

Solution
$$11\overline{)8.^80^30^80^30\ ..}\quad = 0.7272\ ...$$

$$\frac{8}{11} = 0.\dot{7}\dot{2}.$$

0.7272... = $0.\dot{7}\dot{2}$ is a recurring decimal.

Worked Example Writing a recurring decimal as a fraction

Question

Write $0.5\dot{7}$ as a fraction in its simplest form.

> Multiplying by 100 gives a decimal part in (B) the same as that in (A). They disappear when subtracted.

Solution

Suppose $x = 0.5\dot{7}$ (A)　　then　　$100x = 57.5\dot{7}$ (B)

Subtract (B) – (A)　　　　　　　　　$99x = 57$

$$x = \frac{57}{99} = \frac{19}{33}$$

GET IT RIGHT!

Subtracting gets rid of the recurring decimal.

EXAMINER SAYS…

Remember to simplify fractions and check your answer is reasonable.

Worked Example Adding and subtracting fractions

Question

Helen has $5\frac{1}{3}$ metres of fabric. She uses $2\frac{3}{4}$ metres to make a skirt.

What length of fabric is left?

Solution

$$5\frac{1}{3} - 2\frac{3}{4} = 3 + \frac{1}{3} - \frac{3}{4}$$

$$= 3 + \frac{4}{12} - \frac{9}{12}$$ ← Change both fractions to twelfths.

$$= 2 + \frac{16}{12} - \frac{9}{12}$$ ← Change one of the units to twelfths to do the subtraction.

There are $2\frac{7}{12}$ metres of fabric left.

GET IT RIGHT!

To **add** or **subtract** mixed numbers, work out the whole numbers then the fractions.

Worked Example Multiplying and dividing fractions

Question

Work out:

a $\frac{3}{4} \times \frac{2}{5}$ 　　　　　**b** $4\frac{1}{2} \div \frac{3}{4}$

> For × and ÷ you must change any mixed numbers to improper fractions first.

Solution

a $\dfrac{3}{\cancel{4}_2} \times \dfrac{\cancel{2}^1}{5}$

$$= \frac{3}{10}$$

> Simplify before multiplying where possible.

b $\dfrac{9}{2} \div \dfrac{3}{4} = \dfrac{\cancel{9}^3}{\cancel{2}_1} \times \dfrac{\cancel{4}^2}{\cancel{3}_1}$

$$= \frac{6}{1}$$

$$= 6$$

> Only the second fraction is turned upside-down.

> Change improper fractions to whole numbers or mixed numbers at the end.

GET IT RIGHT!

To cancel, one number must be on the top and one on the bottom

Worked Example Simplifying surds

Question

Simplify:

a $\sqrt{3} \times \sqrt{6}$ **b** $2\sqrt{3} + \sqrt{75}$ **c** $\dfrac{\sqrt{21}}{\sqrt{7}}$

Solution

a $\sqrt{3} \times \sqrt{6} = \sqrt{18}$

$\quad = \sqrt{9 \times 2}$

$\quad = 3\sqrt{2}$

b $2\sqrt{3} + \sqrt{75} = 2\sqrt{3} + \sqrt{25 \times 3}$

$\quad = 2\sqrt{3} + 5\sqrt{3}$

$\quad = 7\sqrt{3}$

c $\dfrac{\sqrt{21}}{\sqrt{7}} = \sqrt{\dfrac{21}{7}}$

$\quad = \sqrt{3}$

9 is a square number.

$\sqrt{a} \times \sqrt{b} = \sqrt{ab}$
and $\dfrac{\sqrt{a}}{\sqrt{b}} = \sqrt{\dfrac{a}{b}}$

 **GET IT RIGHT!**

To simplify a surd, make the number under the root as **small** as possible.

 A/A*

THE GRADE To get a grade **A/A*** you must be able to work confidently with surds.

Worked Example Expanding brackets with surds

Question

Write $(5\sqrt{3} - 2)^2$ in the form $a + b\sqrt{3}$ where a and b are integers.

Solution

$(5\sqrt{3} - 2)^2 = (5\sqrt{3} - 2)(5\sqrt{3} - 2)$

$\quad = 25 \times 3 - 10\sqrt{3} - 10\sqrt{3} + 4$

$\quad = 79 - 20\sqrt{3}$ (a is 79 and b is –20)

$\sqrt{3} \times \sqrt{3} = 3$

 **GET IT RIGHT!**

Get it right! You **cannot** just square each term.

 A/A*

THE GRADE To get a grade **A/A*** you must be able to expand brackets that include surds.

Worked Example Rationalising the denominator of a surd

Question

Rationalise the denominators of: **a** $\dfrac{\sqrt{3}}{\sqrt{5}}$ **b** $\dfrac{2}{7 + \sqrt{3}}$

Solution

a $\dfrac{\sqrt{3}}{\sqrt{5}} = \dfrac{\sqrt{3} \times \sqrt{5}}{\sqrt{5} \times \sqrt{5}}$ ← Multiply top and bottom by $\sqrt{5}$

$\quad = \dfrac{\sqrt{15}}{5}$

b $\dfrac{2}{7 + \sqrt{3}} = \dfrac{2}{(7 + \sqrt{3})} \times \dfrac{(7 - \sqrt{3})}{(7 - \sqrt{3})}$ ← Multiply top and bottom by $7 - \sqrt{3}$

$\quad = \dfrac{14 - 2\sqrt{3}}{49 - 7\sqrt{3} + 7\sqrt{3} - 3}$

$\quad = \dfrac{14 - 2\sqrt{3}}{46} = \dfrac{7 - \sqrt{3}}{23}$ ← Divide top and bottom by 2

The $7\sqrt{3}$ terms have disappeared because of the opposite signs.

 **GET IT RIGHT!**

Rationalising the denominator means getting rid of the roots on the bottom.

 **AQA EXAMINER SAYS...**

Always simplify as far as possible.

Decimals, fractions and surds 🎞

1 Work out: **a** $48.52 + 6 \div 10$ **b** $3.2 \times 100 - 0.84 \times 10$

   c $0.4 \times (5 - 1.3)$ **d** $\dfrac{0.2 \times 1.4}{1.2 - 0.85}$

2 Which of the following fractions is nearest to $\frac{3}{5}$? Show how you decide.

$\dfrac{7}{10}$ $\dfrac{1}{2}$ $\dfrac{5}{8}$ $\dfrac{3}{4}$ $\dfrac{13}{20}$

3 Which of the following fractions are equivalent to recurring decimals?

$\dfrac{2}{3}$ $\dfrac{5}{8}$ $\dfrac{5}{9}$ $\dfrac{3}{11}$ $\dfrac{16}{25}$ $\dfrac{5}{6}$ $\dfrac{6}{7}$ $\dfrac{9}{20}$

4 Write these decimals as fractions in their lowest form.

   a $0.\dot{4}$ **b** $0.\dot{3}\dot{7}$ **c** $0.1\dot{2}$ **d** $0.5\dot{4}\dot{9}$

5 Tom has $2\frac{1}{2}$ pints of milk. He drinks $\frac{2}{3}$ pint. How much milk is left?

6 Work out:

   a $\dfrac{1}{3} + \dfrac{1}{6}$ **b** $\dfrac{7}{8} - \dfrac{3}{5}$ **c** $\dfrac{2}{7} \times \dfrac{5}{6}$ **d** $\dfrac{3}{4} \div \dfrac{9}{20}$

   e $3\frac{2}{3} + 1\frac{4}{5}$ **f** $4\frac{1}{2} - 2\frac{5}{8}$ **g** $2\frac{1}{2} \times 2\frac{4}{5}$ **h** $1\frac{3}{4} \div 2\frac{5}{8}$

7 Sally has two cats, Kit and Kat.

Kit eats $\frac{1}{3}$ of a tin of food every day.

Kat eats $\frac{1}{4}$ of a tin of food every day.

What is the **least** number of tins needed
to feed the cats for eight days?

8 Simplify: **a** $\sqrt{5} \times \sqrt{10}$ **b** $\sqrt{27} + \sqrt{12} - 4\sqrt{3}$ **c** $\dfrac{\sqrt{8}}{4}$ **d** $\dfrac{\sqrt{60}}{\sqrt{5}}$

9 Expand and simplify.

   a $(\sqrt{2} + \sqrt{18})^2$ **b** $(\sqrt{5} - 1)^2$

   c $(4 + \sqrt{3})(4 - \sqrt{3})$ **d** $(2 - \sqrt{5})(3 - \sqrt{5})$

10 Rationalise the denominators and simplify where possible.

   a $\dfrac{4}{\sqrt{2}}$ **b** $\dfrac{\sqrt{7}}{\sqrt{3}}$ **c** $\dfrac{3\sqrt{8}}{\sqrt{2}}$

   d $\dfrac{1}{3 + \sqrt{2}}$ **e** $\dfrac{2}{3 - \sqrt{5}}$ **f** $\dfrac{2 - \sqrt{3}}{2 + \sqrt{3}}$

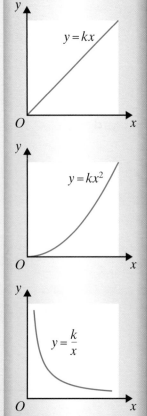

Key points

◎ To **convert a percentage to a fraction or decimal**, divide by 100.

◎ To **convert a fraction or decimal to a percentage**, multiply by 100.

◎ To **calculate compound interest** use a multiplier.
For example, for 5% interest multiply by 1.05 for each year (or add on the interest at the end of each year).

◎ To **write one quantity as a percentage of another**, write them as a fraction, then multiply by 100. (Remember they must be in the **same units**.)

◎ **Percentage increase (decrease)** $= \dfrac{\text{increase (or decrease)}}{\text{original amount}} \times 100$

◎ A **ratio** compares the sizes of two or more quantities or numbers.
A **proportion** compares one part with the whole.
For example, with 11 boys and 13 girls, the **ratio** of boys to girls is $11:13$
The **proportion** of boys is $\frac{11}{24}$ and the **proportion** of girls is $\frac{13}{24}$.

◎ To **simplify a ratio** divide both (all) numbers by the same number or multiply them by the same number. (Make sure you put the parts in the right order and the same units first.)

◎ y **is directly proportional to** x can be written as $y \propto x$
$y \propto x$ means $y = kx$ where x and y are variables and k is a constant.
The graph of y against x is a **straight line through the origin, (0, 0)** with **gradient** k.
y **is directly proportional to** x^2 (or $y \propto x^2$) means $y = kx^2$
y **is directly proportional to** x^3 (or $y \propto x^3$) means $y = kx^3$
y **is directly proportional to** $\sqrt{x}$ (or $y \propto \sqrt{x}$) means $y = k\sqrt{x}$ etc.

◎ y **is inversely proportional to** x can be written as $y \propto \dfrac{1}{x}$
$y \propto \dfrac{1}{x}$ means $y = \dfrac{k}{x}$ where x and y are variables and k is a constant.

y **is inversely proportional to** x^2 (or $y \propto \dfrac{1}{x^2}$) means $y = \dfrac{k}{x^2}$

y **is inversely proportional to** x^3 (or $y \propto \dfrac{1}{x^3}$) means $y = \dfrac{k}{x^3}$

y **is inversely proportional to** $\sqrt{x}$ (or $y \propto \dfrac{1}{\sqrt{x}}$) means $y = \dfrac{k}{\sqrt{x}}$ etc.

Worked Example — Increasing by a percentage

Question

A builder charges £240 plus VAT at $17\frac{1}{2}$ % for laying a path. What is the total bill?

Solution

Read the question carefully.
Does it ask for the VAT or the total?

Without a calculator	With a calculator
10% = £240 ÷ 10 = £24 5% = $\frac{1}{2}$ of 10% = £12 2.5% = $\frac{1}{2}$ of 5% = £6 ──────── 17.5% (VAT) = £42 Total bill = £240 + £42 = £282	17.5% = £240 ÷ 100 × 17.5 = £42 (or 0.175 × £240 = £42) Total bill = £42 + £240 = £282 **Alternative method:** Total bill = 1.175 × £240 = £282 <small>100% + 17.5% = 117.5% = 1.175</small>

(beside the first table: 17.5% = 10% + 5% + 2.5%)

Worked Example — Decreasing by a percentage

Question

A new car costs £18 000.
Its value falls by 25% in the first year and 15% in the second year.

What is the value of the car after 2 years?

Solution

£18 000 × 0.75 × 0.85 = £11 475

EXAMINER SAYS...

Using a multiplier is the most efficient method for this type of question.

Worked Example — Compound interest

Question

Sharon invests £3000 at 5.4% compound interest.

How much will she have in the account at the end of 4 years?

Read the question carefully.
Does it ask for the interest or the amount?

Solution

	Amount and beginning of year	Amount at end of year
Year 1	£3000	£3000 × 1.054 = £3162
Year 2	£3162	£3162 × 1.054 = £3332.748
Year 3	£3332.748	£3332.748 × 1.054 = £3512.716392
Year 4	£3512.716392	£3512.716392 × 1.054 = £3702.403077

The amount at the end of 4 years = £3702.40

<small>The interest earned = £3702.40 − £3000 = £702.40</small>

Alternatively, the amount at the end of 4 years = £3000 × 1.054^4

= £3702.40 (nearest penny)

AQA

EXAMINER SAYS...

Money answers should be rounded to the **nearest penny**.

Worked Example — Percentage increase or decrease

Question

A shopkeeper buys computer games for £15 each and sells them for £19.95 each.

What is the percentage profit?

On a calculator:

$4.95 \div 15 = 0.33 = 33\%$

EXAMINER SAYS...

Candidates often lose marks by using the wrong amount in the denominator

Solution

$$\text{Profit} = £19.95 - £15 = £4.95$$

GET IT RIGHT!

The denominator must be the **original** amount

Working in pounds:

$$\text{Percentage profit} = \frac{4.95}{15} \times \frac{100}{1} = \frac{495^{99}}{15_3}$$

$$\text{Percentage profit} = 33\%$$

Working in pence:

$$\text{Percentage profit} = \frac{495}{1500} \times \frac{100}{1} = \frac{495^{99}}{15_3}$$

$$\text{Percentage profit} = 33\%$$

Worked Example — Reversing a percentage change

EXAMINER SAYS...

Check your answer by working backwards.

Question

In a sale, the price of a computer is reduced by 30% to £840.

What was the original price?

Solution

70% of the original price	= £840	
1% of the original price	= £840 ÷ 70	= £12
100% of the original price	= £12 × 100	= £1200

GET IT RIGHT!

For a **decrease, subtract from 100%**
100% − 30% = 70%
For an **increase, add to 100%**
100% + 30% = 130%

Worked Example — Using ratios

Question

When Ann, Bob and Carol set up a business Ann invested £2500, Bob invested £3750 and Carol invested £6250. In the first year the business makes a profit of £1600.

Divide the profit between Ann, Bob and Carol so their shares are in the same ratio as the amounts they invested.

Solution

Ann's share : Bob's share : Carol's share $= 2500 : 3750 : 6250 = 250 : 375 : 625$

$$= 10 : 15 : 25 = 2 : 3 : 5$$

This means Ann gets $\frac{2}{10}$, Bob gets $\frac{3}{10}$ and Carol gets $\frac{5}{10}$ of the profit.

1 part $= \frac{1}{10}$ of £1600 = £160

Ann's share = £160 × 2 = £320

The profit is split into 10 equal parts. Ann gets 2, Bob gets 3 and Carol gets 5

Bob's share = £160 × 3 = £480

Carol's share = £160 × 5 = £800

Check that the total is £1600

Worked Example   Using ratios in different ways

Question   The recipe for a drink says 'Mix 2 parts pineapple juice with 3 parts orange juice'.

a How much pineapple juice should you mix with 600 ml of orange juice?

b How much pineapple juice do you need to make 1.5 litres of the drink?

Solution

a 3 parts orange juice   = 600 ml, so 1 part = 200 ml

2 parts pineapple juice = 400 ml.

There are 600 ml of orange juice and the recipe says orange juice is 3 parts. Find 1 part, then the quantity needed.

b 5 parts = 1.5 litres = 1500 ml ←——— 1.5 litres is the total amount, so this is 5 parts.

1 part = 300 ml

2 parts pineapple juice = 600 ml

GET IT RIGHT!

Make sure you use the right parts of the ratio.

Worked Example   Direct and inverse proportion

Question y is inversely proportional to x, and $y = 10$ when $x = 4$.

a Find the value of y when $x = 20$.

b Sketch the graph of y against x for positive values of x

Solution

a y is inversely proportional to x means $y = \dfrac{k}{x}$

Substituting the given values: $10 = \dfrac{k}{4}$

Multiplying by 4 $40 = k$

The equation for y in terms of x is $y = \dfrac{40}{x}$

When $x = 20$, $y = \dfrac{40}{20} = 2$

When x is multiplied by 5, y is divided by 5.

GET IT RIGHT!

Don't confuse direct and inverse proportionality. y is **directly** proportional to x means $y = kx$ (The graph would be a straight line through O.)

b

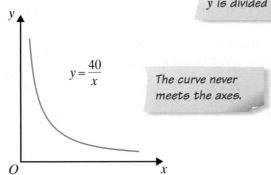

$y = \dfrac{40}{x}$

The curve never meets the axes.

THE GRADE   To get a grade **A** you must be able to do questions involving inverse proportion.

Percentages and proportion

Time Yourself!

Can you complete these questions in **40** minutes?

1 A small tin of paint contains 20% less than a large tin.
The large tin contains 2.5 litres. How much paint is in the small tin?

2 A game costs £26.80 before VAT is added at a rate of 17.5%. What is the price including VAT?

3 Andy earns £366 per week. He is awarded a pay rise of 4.3%.
How much does he earn each week after the pay rise?

4 Mrs Young buys a car for £40 000.
It decreases in value at the rate of 25% each year.
Find the value of the car after 2 years.

5 Sally invests £6000 for 5 years in a savings account that pays 5% compound interest per year.
How much will she have in the account at the end of 5 years?

6 A school buys 10 bottles of milk for drinks at a parents' evening.
Each bottle holds enough for 25 drinks. The number of drinks needed is 235.
Calculate the percentage of milk that will be used.

7 A house is valued at £275 000. A year later it is valued at £319 000.
What is the percentage increase in value?

8 Moira sees this sign in a shop window.
How much were the boots before the price reduction?

PRICE REDUCTION
BOOTS 35% OFF
Now £31.20

9 Ruth, Sue and Tom serve meals at a café. Each week they share their tips in the ratio of the numbers of hours they work. One week when Ruth works for 12 hours, Sue works for 18 hours and Tom works for 20 hours, they get £140 in tips. How much does each person get?

10 The table shows measurements of t and v.
Which of these relationships is true?

t	1	2	5
v	0.5	4	62.5

a $v \propto t$ **b** $v \propto t^3$ **c** $v \propto \frac{1}{t}$

You **must** show your working.

11 Y and Z are both positive quantities. Y is directly proportional to Z^2.
When Y is 72, $Z = 3$.

a Express Y in terms of Z.   **b** What is the value of Y when Z is 5?

c What is the value of Z when Y is 18?

12 P and Q are both positive quantities. P is inversely proportional to the square root of Q. When P is 6, $Q = 4$.

Find the value of P when $Q = 16$.

Indices and standard form

Key words

square number

cube number

square root

cube root

index, power or
exponent

indices

standard form

Key points

⊚ You should know these **square** numbers (and the related square roots):

$2^2 = 4, 3^2 = 9, 4^2 = 16, 5^2 = 25, 6^2 = 36, 7^2 = 49, 8^2 = 64, 9^2 = 81,$
$10^2 = 100, 11^2 = 121, 12^2 = 144, 13^2 = 169, 14^2 = 196, 15^2 = 225$

⊚ You should know these **cube** numbers (and the related cube roots):

$2^3 = 8, 3^3 = 27, 4^3 = 64, 5^3 = 125, 10^3 = 1000$

⊚ An **index**, **power** or **exponent** tells you how many times the base number is multiplied by itself.

For example,　　$2^5 = 2 \times 2 \times 2 \times 2 \times 2 = 32$

5 is the index

2 is the base number

The plural of index is **indices**.

⊚ On a calculator:

x^2　　$\sqrt{}$　　x^3　　$\sqrt[3]{}$　　x^y

squares　square roots　cubes　cube roots　indices

⊚ Rules of indices:

To **multiply**, **add the indices**, for example, $5^3 \times 5^4 = 5^7$
To **divide**, **subtract the indices**, for example, $2^5 \div 2^2 = 2^3$
To **find powers of powers**, write out (or multiply indices), for example, $(7^3)^2 = 7^3 \times 7^3 = 7^6$
A **negative index** means '**1 over**', for example,

$2^{-3} = \dfrac{1}{2^3} = \dfrac{1}{8}$

A **zero index** always gives **1**, for example, $5^0 = 1$
A **fraction index** gives a **root**, for example, $9^{\frac{1}{2}} = \sqrt{9} = 3, 8^{\frac{1}{3}} = \sqrt[3]{8} = 2$

⊚ **Standard form** is a shorthand way of writing very big and very small numbers.

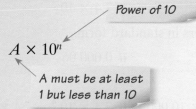

Power of 10

$A \times 10^n$

A must be at least
1 but less than 10

⊚ Input 2.7×10^8 on your calculator as 2.7 [EXP] 8 or 2.7 [EE] 8

Worked Example Powers and roots

Question

Find the value of

a $(-3)^2$ **b** -3^2 **c** 3^{-2} **d** 2^0 **e** $36^{\frac{1}{2}}$ **f** $8^{\frac{2}{3}}$ **g** $25^{-\frac{3}{2}}$

– in the power means '1 over'.

Solution

a $(-3)^2 = -3 \times -3 = 9$ **b** $-3^2 = -3 \times 3 = -9$ **c** $3^{-2} = \dfrac{1}{3^2} = \dfrac{1}{9}$

d $2^0 = 1$ **e** $36^{\frac{1}{2}} = \sqrt{36} = 6$

f $8^{\frac{2}{3}} = (\sqrt[3]{8})^2$ or $\sqrt[3]{8^2} = \sqrt[3]{64}$

$= 2^2$ $= 4$ *but it is easier to take the cube root first*

$= 4$

g $25^{-\frac{3}{2}} = \dfrac{1}{(\sqrt{25})^3}$

$= \dfrac{1}{5^3} = \dfrac{1}{125}$

GET IT RIGHT!

Take care!
For example, 3^{-2}
does not mean
3×-2 or $3 - 2$

A/A*
THE GRADE

To get a grade **A/A*** you must be able to work with fractional powers (indices).

Worked Example Rules of indices

Question

Simplify **a** $\dfrac{2^3}{2^4}$ **b** $\dfrac{3^2}{2^4}$ **c** $\dfrac{x^6 \times x^4}{x^{12}}$ **d** $(a \times a^4)^3$

Solution

a $\dfrac{2^3}{2^4} = 2^{-1}$ **b** $\dfrac{3^2}{2^4} = \dfrac{3 \times 3}{2 \times 2 \times 2 \times 2}$

Or cancel:

$= \dfrac{1}{2}$ $\dfrac{2 \times 2 \times 2}{2 \times 2 \times 2 \times 2}$ $= \dfrac{9}{16}$

x^{-2} is equal to $\dfrac{1}{x^2}$

Remember $a = a^1$

GET IT RIGHT!

You can only
add or subtract
powers if the
**base numbers
(or letters) are
the same**

c $\dfrac{x^6 \times x^4}{x^{12}} = \dfrac{x^{10}}{x^{12}} = x^{-2}$ **d** $(a \times a^4)^3 = (a^5)^3 = a^{15}$

To multiply, <u>add</u> indices.
To divide, <u>subtract</u> indices.

To <u>find powers of powers</u>, <u>multiply indices</u>, or write $(a^5)^3 = a^5 \times a^5 \times a^5 = a^{15}$

Worked Example Standard form numbers and ordinary numbers

Question

a Write these standard form numbers as ordinary numbers.

i 2.3×10^4 **ii** 7.4×10^{-3}

b Write these numbers in standard form.

i 84 500 000 **ii** 0.000 06

Solution

a i $2.3 \times 10^4 = 2.3 \times 10\ 000$ **ii** $7.4 \times 10^{-3} = 7.4 \times 0.001$

$= 23\ 000$ $= 0.0074$

Multiplying by 10^{-3} is the same as dividing by 10^3

b i $84\ 500\ 000 = 8.45 \times 10^7$ **ii** $0.000\ 06 = 6 \times 10^{-5}$

Worked Example Calculating in standard form

Question

Work these out. Give your answers in standard form.

a $(4 \times 10^3) \times (5 \times 10^7)$ **b** $(4 \times 10^3) \div (5 \times 10^7)$

Solution

GET IT RIGHT!

The first part, *A*, **must** be between 1 and 10.

a $(4 \times 10^3) \times (5 \times 10^7)$

$= 20 \times 10^{10}$

$= 2 \times 10 \times 10^{10}$

$= 2 \times 10^{11}$

Multiply the ordinary numbers, then add the powers of 10

b $(4 \times 10^3) \div (5 \times 10^7)$

$= 0.8 \times 10^{-4}$

$= 8 \times 10^{-1} \times 10^{-4}$

$= 8 \times 10^{-5}$

Divide the ordinary numbers, then subtract the powers of 10

Indices and standard form

END OF CHAPTER QUESTIONS

Time Yourself!

Can you complete these questions in **40** minutes?

1 Find the value of: **a** 7^3 **b** 6^0 **c** 2^{-4} **d** -2^4 **e** $(-2)^4$
      f $16^{\frac{1}{2}}$ **g** $27^{\frac{2}{3}}$ **h** $36^{-\frac{1}{2}}$ **i** $4^{-\frac{3}{2}}$ **j** $125^{-\frac{2}{3}}$

2 Find the value of: **a** $2^4 \times 3^2$ **b** $10^3 \div 10^5$ **c** $(2^2)^3$ **d** $\dfrac{2^3}{3^2}$

3 Simplify: **a** $x^5 \times x^4$ **b** $x^8 \div x^2$ **c** $(x^6)^3$ **d** $\dfrac{x^3 \times x}{x^7}$

4 Write these numbers in standard form.
   a 9 000 000   **b** 0.000 04   **c** 462 000 000 000   **d** 0.000 000 057

5 Write these standard form numbers as ordinary numbers.
   a 3×10^8 **b** 7.42×10^{10} **c** 3×10^{-4} **d** 2.56×10^{-9}

6 Work these out. Give your answers in standard form.
   a $(5 \times 10^4) \times (8 \times 10^7)$ **b** $(3 \times 10^{12}) \div (4 \times 10^5)$ **c** $(9 \times 10^{-4})^2$
   d Half of 6×10^4 **e** 15% of (3×10^6) *Use your calculator to check your answers.*

7 Work these out on your calculator. Give your answer in standard form to 3 significant figures.
   a $(3.28 \times 10^7) \times (9.67 \times 10^{-3})$ **b** $(8.39 \times 10^4) \div (1.76 \times 10^{-8})$
   c $(6.25 \times 10^5)^{-3}$ **d** $(4.62 \times 10^{-5}) + (1.97 \times 10^{-6})$

8 Approximately 60 million people live in the UK and their total wealth is approximately £(4×10^{12})
On average, how much is this per person?
Give your answer to 1 significant figure.

AQA EXAMINER SAYS…

Practice standard form questions with and without a calculator for the calculator and non-calculator papers.

1 (a) Write 560 as the product of its prime factors. Give your answer in index form. *(3 marks)*

(b) You are given that $36 = 2^2 \times 3^2$

Write each of these numbers as the product of prime factors in index form.

    (i) 72 *(1 mark)*   (ii) 360 *(1 mark)*

(c) Work out the highest common factor (HCF) of 360 and 560. *(2 marks)*

2 Use approximations to estimate the value of $\dfrac{318 \times 5.09}{0.395}$

You **must** show your working. *(3 marks)*

3 A girl runs 60 metres in 12 seconds.

Both values are measured to an accuracy of two significant figures.

What is her least possible average speed? *(3 marks)*

4 (a) Prove that $0.\dot{4}\dot{6} = \frac{46}{99}$ *(2 marks)*

(b) Hence, or otherwise, express $0.1\dot{4}\dot{6}$ as a fraction. *(2 marks)*

5 Rationalise the denominator of $\dfrac{2 + \sqrt{6}}{\sqrt{6}}$. Simplify your answer fully. *(3 marks)*

6 Two rectangles, A and B, are equal in area.

$(\sqrt{15} + \sqrt{3})$cm

$(\sqrt{15} - \sqrt{3})$cm   A $\sqrt{2}$ cm   B   Not to scale

Calculate the length of rectangle B. Give your answer in the form $a\sqrt{2}$. *(4 marks)*

7 (a) Alice sees two advertisements for the same MP3 player.

The rate of VAT is $17\frac{1}{2}\%$

> **The Audio Store**
> **£92**
> Price includes VAT

> **CUT PRICES**
> **£80**
> Price does **NOT** include VAT

Alice works out that the MP3 player costs more at Cut Prices than at The Audio Store.

How much more does it cost at Cut Prices? *(3 marks)*

(b) In a sale the price of a digital radio decreases from £75 to £66.

Work out the percentage decrease in price. *(3 marks)*

(c) In a sale the price of a CD player decreases by 60%. The sale price is £18.60

Work out the price before the sale. *(3 marks)*

AQA Spec B, Intermediate Module 3, Nov 06

8 £4200 is invested at 3.5% compound interest per annum.

How many years will it take for the investment to exceed £5000? *(3 marks)*

9 A box of chocolates contains 18 milk chocolates and 12 dark chocolates.

A larger box containing 40 chocolates has the same ratio of milk to dark chocolates.

How many dark chocolates are in the larger box? *(3 marks)*

10 In an experiment measurements of t and w were taken.

These are the results. Which of these rules fits the results?

t	2	3	6
w	6	13.5	54

(a) $w \propto t$ (b) $w \propto t^2$ (c) $w \propto t^3$ You **must** show all your working. *(4 marks)*

11 W and D are both positive quantities. W is inversely proportional to D.

When W is 25, $D = 16$. Find the value of W when $D = W$. *(4 marks)*

12 (a) (i) Evaluate $6p^0$ (ii) Evaluate $(6p)^0$ *(2 marks)*

(b) If $2^x = \frac{1}{16}$ find the value of x. *(2 marks)*

(c) If $27^y = 81^{\frac{1}{2}}$ find the value of y. *(2 marks)*

13 Use your calculator to work out:

(a) the reciprocal of 0.8 *(1 mark)*

(b) $\sqrt{6.4^2 + 3.18^3}$

(i) Write down the full calculator display. *(1 mark)*

(ii) Write your answer to 2 significant figures. *(1 mark)*

AQA Spec B, Higher Module 3, Nov 06

14 The Moon is approximately a sphere of radius 3480 km.

The surface area of a sphere is given by the formula $A = 4\pi r^2$

Calculate the approximate surface area of the Moon.

Give your answer in standard form. *(3 marks)*

1   The area of this rectangle is 24 cm².

Find the value of *x*, writing your answer in

the form $a\sqrt{b}$ where *a* and *b* are integers.

x cm

$4\sqrt{3}$ cm

$$x = \frac{24}{4\sqrt{3}} = \frac{6}{\sqrt{3}} = \frac{6 \times \sqrt{3}}{\sqrt{3} \times \sqrt{3}} = \frac{6\sqrt{3}}{3}$$

Attempting to rationalise the denominator
gives the second mark

The candidate gets the
first mark for using
width = area ÷ length.

Answer ....$2\sqrt{3}$..................

The third mark is awarded
for the correct final answer

(3 marks)

 2   A plumber's bill is £376 including $17\frac{1}{2}$% VAT.

What was the price charged without VAT?

The first mark
is awarded
for 117.5% or
1.175

117.5% of the price charged before VAT = £376

1% of the price charged before VAT = £376 ÷ 117.5 = £3.2

100% of the price charged before VAT = £3.20 × 100

Dividing by
117.5 and
multiplying by
100 (or dividing
by 1.175) gains
the second
mark.

The candidate is awarded the
third mark for the correct value.

£320

Answer ....

(3 marks)

3   The planet Mars is approximately a sphere of radius 3400 km.

The volume of a sphere is given by the formula $V = \frac{4}{3}\pi r^3$.

Calculate the approximate volume of the planet Mars.

Give your answer in standard form to an appropriate degree of accuracy.

$$V = \frac{4}{3}\pi r^3 = \frac{4}{3}\pi \times 3400^3$$

The candidate is awarded one mark for
substituting 3400 correctly into the formula.

$$V = 1.6463621 \times 10^{11}$$

The candidate is awarded another mark for
calculating the value correctly.

The final mark is for rounding.

1.6×10^{11}

Answer ....% .

(3 marks)

AQA
EXAMINER SAYS...

The value of π on your calculator is
the most accurate but other values
would be accepted.

AQA
EXAMINER SAYS...

Rounding to 1 s.f., 2 s.f., 3 s.f. or 4 s.f.
would be accepted to get the final mark.

The second mark is for finding the value of $27^{\frac{1}{3}}$

4   Find the value of $2^{-3} \times 27^{\frac{1}{3}}$

$$2^{-3} = \frac{1}{2^3} = \frac{1}{8}$$

$$27^{\frac{1}{3}} = \sqrt[3]{27} = 3$$

$$2^{-3} \times 27 = \frac{1}{8} \times 3$$

The candidate is awarded one
mark for finding the value of 2^{-3}

Answer ....$\frac{3}{8}$..................

The final mark is awarded for
multiplying correctly.

(3 marks)

Area and volume

area

perimeter

volume

rhombus

parallelogram

trapezium

kite

cube

cuboid

prism

cylinder

sphere

hemisphere

pyramid

cone

frustum
(of a cone)

circle

circumference

arc (of a circle)

sector (of a circle)

segment

To find the arc length work out what fraction of the circumference you need.

To find the sector area work out what fraction of the area you need.

Key points

Areas

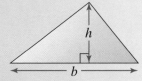

⊚ Area of **triangle** = $\frac{1}{2}bh$

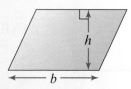

⊚ Area of **parallelogram** = bh

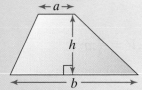

⊚ Area of **trapezium** = $\frac{1}{2}(a + b)h$

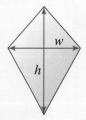

⊚ Area of **kite** = $\frac{1}{2}hw$

Circles

⊚ **Circumference** = $\pi d = 2\pi r$

⊚ **Area** = πr^2

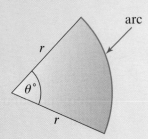

⊚ **Arc length** = $\frac{\theta}{360} \times \pi d$

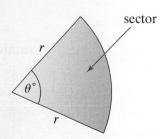

⊚ **Area of sector** = $\frac{\theta}{360} \times \pi r^2$

3-D solids

Shape	Volume	Surface area

Prism

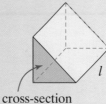

cross-section

V = area of cross-section × length

SA = sum of areas of all the faces

> In this triangular prism there are 2 triangular faces, 1 large rectangular face and 2 smaller rectangular faces.

Cylinder

$V = \pi r^2 h$

$SA = \pi dh + 2\pi r^2$
where $d = 2r$

Sphere

$V = \frac{4}{3}\pi r^3$

$SA = 4\pi r^2$

Cone

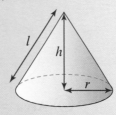

$V = \frac{1}{3}\pi r^2 h$

$SA = \pi rl + \pi r^2$

Pyramid

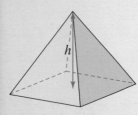

$V = \frac{1}{3}$ × base area × perpendicular height

SA = sum of areas of all the faces

> In this square-based pyramid there are 4 congruent triangular faces and 1 square face.

Worked Example Area and perimeter

Question

The diagram shows an area of wooden decking alongside a semicircular pond.

a Calculate the perimeter of the decking.

b Calculate the area of wood used in the decking.

Give your answers in terms of π.

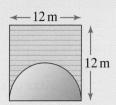

Solution

Perimeter is a length so choose the right units

a The perimeter of the decking is 3 sides of the square and the circumference of the semicircle.

Perimeter $= 12 + 12 + 12 + \dfrac{\pi d}{2} = 36 + \dfrac{\pi \times 12}{2}$ ← Divide by 2 for a semicircle.

$= 36 + 6\pi$ m

b The area of the decking is the area of the rectangle – the area of the semicircle.

Area of semicircle $= \dfrac{\pi r^2}{2} = \dfrac{\pi \times 6^2}{2} = 18\pi$ m^2 ← Use the radius, not the diameter.

Area of decking $= 12 \times 12 - 18\pi = (144 - 18\pi)$ m^2

THE GRADE To get a grade **A/A*** you must be confident working with sectors, segments and arc lengths so practise the following type of question.

πr^2 means $\pi \times r \times r$
not $(\pi r)^2$ or $\pi r \times 2$

AQA EXAMINER SAYS...

The question asks you to leave your answer in terms of π. Don't try to simplify further, $144 - 18\pi \neq 126\pi$

Worked Example Arc length, area of sectors and segments

Question

A circle has a radius of 2.8 cm. A sector has an arc length of 4.4 cm.

The angle at the centre of the sector is θ.

a Calculate the value of θ.

b Calculate the area of the shaded segment.

2.8 cm

4.4 cm

An arc is a fraction of the circumference so use $C = \pi d$.

a Arc length $= \dfrac{\theta}{360} \times \pi d$ ← Use the diameter, not the radius.

So $\quad 4.4 = \dfrac{\theta}{360} \times \pi \times 5.6$ ← Multiply both sides by 360

$360 \times 4.4 = \theta \times \pi \times 5.6$ ← Divide both sides by $\pi \times 5.6$

$\theta = \dfrac{360 \times 4.4}{\pi \times 5.6} = 90.0°$ (1 d.p.)

Solution

A sector is a fraction of the area so use $A = \pi r^2$.

b Area of shaded segment = area of sector – area of triangle

Area of sector $= \dfrac{\theta}{360} \times \pi r^2$

$= \dfrac{90}{360} \times \pi \times 2.8^2$

$= 6.157... = 6.16$ cm^2 (3 s.f.)

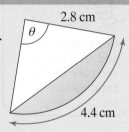

AQA EXAMINER SAYS...

Remember not to round your answers too early. Use all of the numbers on your calculator display and round your final answer.

51

$$\text{Area of triangle} = \frac{1}{2} \times \text{base} \times \text{height}$$

θ = 90° so it is a right-angled triangle and you can use this formula. The base and height are both radii.

$$= \frac{1}{2} \times 2.8 \times 2.8$$

$$= 3.92 \text{ cm}^2$$

Remember to include the correct units.

$$\text{Area of shaded segment} = 6.16 - 3.92 = 2.24 \text{ cm}^2 \text{ (3 s.f.)}$$

Worked Example Volume and surface area

Question

The area of the base of a cylinder is 9π cm².

The volume of the cylinder is 36π cm³.

Work out the **total** surface area of the cylinder.

Solution

Volume of a cylinder = base area × height

A cylinder is a prism.

So
$$36\pi = 9\pi \times h$$

$$h = \frac{36\pi}{9\pi} = 4 \text{ cm}$$

Divide by 9π

Base area of a cylinder = πr^2

So
$$9\pi = \pi r^2$$

$$r^2 = \frac{9\pi}{\pi} = 9$$

Divide by π

$$r = 3 \text{ cm}$$

Find the square root.

$$d = 2r = 6 \text{ cm}$$

GET IT RIGHT!

Remember that the curved face is a rectangle and its area = circumference of base × height of cylinder.

GET IT RIGHT!

Don't forget to add on the top and base for the **total** surface area.

Total surface area of cylinder = ⬤ Area of top + ▭ Area of curved face + ⬤ Area of base

6 cm

4 cm

$$= 9\pi + \pi dh + 9\pi$$

$$= 9\pi + \pi \times 6 \times 4 + 9\pi$$

$$= 9\pi + 24\pi + 9\pi$$

$$= 42\pi \text{ cm}^2$$

$$= 132 \text{ cm}^2 \text{ (3 s.f.)}$$

You were given this area in the question. Don't waste time working it out again.

Work in terms of π to avoid rounding errors.

Remember to include the correct units.

BUMP UP
A/A*
THE GRADE To get a grade **A** you need to be able to find the volume and surface area of pyramids, cones and spheres. For an **A***, you might be asked to find the volume of the frustum of a cone.

EXAMINER SAYS...

On the non-calculator paper you may be asked to leave your answer in terms of π.

Area and volume

1 a This T-shape is
made of rectangles.

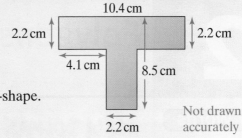

10.4 cm

2.2 cm   2.2 cm

4.1 cm   8.5 cm

2.2 cm

Not drawn
accurately

Calculate the area of the T-shape.

b This shape is made of two semicircles.

The radius of the larger semicircle is 6 cm.

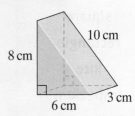

6 cm

Calculate the perimeter of the shape.

2 The diagram shows the dimensions of a
piece of cheese in the shape of a triangular prism.

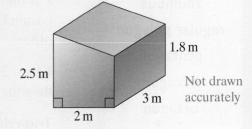

10 cm

8 cm

6 cm   3 cm

a Calculate the volume of the cheese.

b Calculate the amount of shrink-wrap needed
to cover the cheese.

3 a The grey shaded part of the diagram
shows the side wall of a shed.
Calculate the area of the side wall.
State the units of your answer.

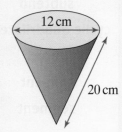

1.8 m

2.5 m

3 m

2 m

Not drawn
accurately

b The shed is a prism with a uniform
cross-section.
Calculate the volume of the shed.

4 A cuboid is made from centimetre cubes. The area of the base is 7 cm^2.
The volume of the cuboid is 14 cm^3. Work out the surface area of the
cuboid.

5 Calculate the **total** surface area of the child's spinning top.
Give your answer in terms of π.
State the units of your answer.

12 cm

20 cm

6 The surface area of a spherical ball-bearing is 222 mm^2.
Calculate the radius of the ball-bearing.

7 A sphere of radius 3 cm has the same volume as a cone of
perpendicular height 3 cm.
Calculate the base radius, r, of the cone.

8 AOB is a sector of a circle of radius 8 cm.
The area of the minor sector AOB is 37 cm^2.
Calculate the size of angle AOB.

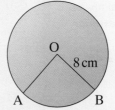

O
8 cm
A   B

Not drawn
accurately

Polygons and circles

Key points

Quadrilaterals

- **Square** – 4 equal sides, 4 right angles, 2 pairs of parallel sides, equal diagonals that bisect at right angles.

- **Rectangle** – 2 pairs of equal and parallel sides, 4 right angles, equal diagonals that bisect each other.

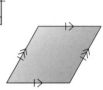

- **Rhombus** – 4 equal sides, 2 pairs of parallel sides, diagonals bisect at right angles.

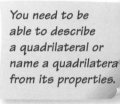

- **Parallelogram** – 2 pairs of equal and parallel sides, diagonals bisect each other.

You need to be able to describe a quadrilateral or name a quadrilateral from its properties.

- **Kite** – 2 pairs of equal adjacent sides, long diagonal bisects shorter one at right angles.

- **Trapezium** – 1 pair of parallel sides.

- **Isosceles trapezium** – 1 pair of parallel sides, other pair equal, diagonals equal.

Polygons

- **Regular polygon** – all sides and angles are equal.

- **Exterior angles** of a polygon add up to 360°.

- **Interior angles** of a polygon add up to $(n - 2) \times 180°$, where n is the number of sides.

  exterior angle + interior angle = 180°

For example, a pentagon has 5 sides so the sum of its interior angles = $(5 - 2) \times 180° = 540°$

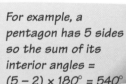

Circles

- The **angle** subtended by an arc **at the centre** of a circle is **twice** the **angle** subtended **at the circumference**.
  angle AOB = 2 × angle ACB

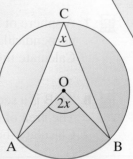

For example, a regular octagon has 8 equal sides and equal angles so each exterior angle = $\frac{360°}{8} = 45°$

The **angle in a semicircle** is a **right angle**.

angle PRQ = 90°

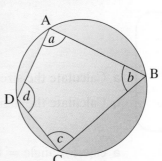

POQ is a diameter.

The **opposite angles** of a **cyclic quadrilateral** add up to **180°**.

$a + c = 180°, b + d = 180°$

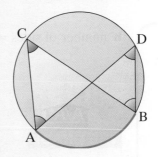

Angles subtended by the **same arc** are **equal**.

angle ACB = angle ADB

angle CAD = angle CBD

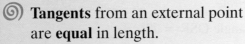

You need to be able to prove the angle and tangent/chord properties of a circle.

The **tangent** to a circle is **perpendicular** to the **radius** at that point.

angle PQO = angle PRO = 90°

Tangents from an external point are **equal** in length.

PQ = PR

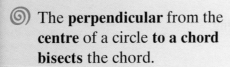

PS and PT are tangents, OQ and OR are radii.

The **perpendicular** from the **centre** of a circle **to a chord** **bisects** the chord.

AX = XB

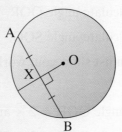

The **alternate segment theorem** says that the angle between a **tangent** and **chord** is **equal** to the **angle** in the **alternate segment**.

angle BAS = angle ACB
angle CAT = angle ABC

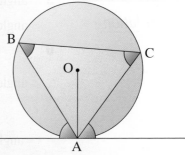

Worked Example Polygons

Question

The diagram shows part of a regular polygon.

Each interior angle is 162°.

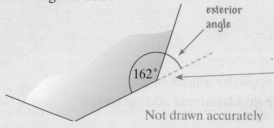

exterior angle

162°

Not drawn accurately

It is a good idea to use the diagram to help you. Add the line to see where the exterior angle is.

GET IT RIGHT!

Remember that the exterior angles of a polygon add up to 360° (not 180°).

a Calculate the size of an exterior angle of the polygon.

b Calculate the number of sides of the polygon.

Solution

a exterior angle = 180° − 162° = 18°   The exterior and interior angle are on a straight line so together they make 180°.

GET IT RIGHT!

This example would be on the non-calculator paper so cancel your answer carefully.

b number of sides = $\frac{360}{18} = \frac{60}{3} = 20$

Exterior angles add up to 360°, so there must be $\frac{360}{18}$ exterior angles and sides.

A/A*

THE GRADE   To get a grade **A/A*** you must be able to use and prove the alternate segment theorem so make sure you understand the following type of question.

Worked Example Circle properties

Question

Q, R and S are points on the circumference of a circle centre O.

ST is a tangent to the circle.

Angle RST = 68° and angle QSR = 72°.

a Calculate angle QOR.

b Calculate angle SQO.

You **must** show your working.

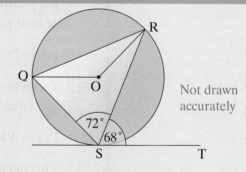

R

Q

O

Not drawn accurately

72°

68°

S   T

AQA

EXAMINER SAYS...

'You **must** show your working' means that you will not get full marks if you do not show your method.

Solution

a   angle QOR = 2 × angle QSR (angle at centre twice angle at circumference)

angle QOR = 2 × 72° = 144°

b   QO = OR (radii)

angle OQR = $\frac{180° - \text{angle QOR}}{2}$ (base angle, isosceles triangle)

= $\frac{180 - 144°}{2} = \frac{36°}{2} = 18°$

angle SQR = angle RST = 68° (alternate segment)

angle SQO = angle SQR − angle OQR

= 68° − 18° = 50°

GET IT RIGHT!

Angle SQR is not 90° and angle SQO is not half of angle SQR. There are no parallel lines so no alternate angles.

Polygons and circles

1 The diagram shows a quadrilateral ABCD.

Angle A = 72°, angle B = 150° and angle C = 88°.

Calculate the exterior angle at D, marked x on the diagram.

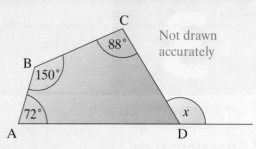

Not drawn accurately

2 P, Q, and S are points on the circumference of a circle, centre O.

QR and SR are tangents.

Angle QRS = 50°.

Giving reasons for your answers, work out

a the value of x **b** the value of y.

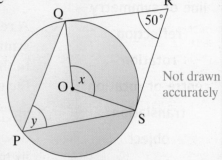

Not drawn accurately

3 A, B and C are points on the circumference of a circle, centre O.

CD is a tangent to the circle.
Angle AOC = 120° and angle BCD = 75°.
Work out the value of x.
You **must** show your working.

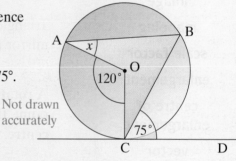

Not drawn accurately

4 Prove the following circle properties.

a The angle subtended at the circumference by a semicircle is a right angle.

b The angle between the tangent and a chord is equal to the angle in the alternate segment.

5 a O is the centre of the circle

Prove that angle a = angle b

b Prove that ABCD is a cyclic quadrilateral.

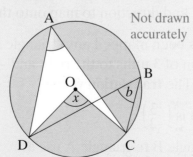

Not drawn accurately

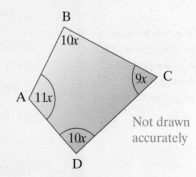

Not drawn accurately

Transformations

Key words

transformation
line of symmetry
reflection
rotation
centre of rotation
translation
object
image
mapping
scale factor
enlargement
centre of enlargement
vector
scalar
collinear

Key points

◎ A **transformation** changes the **position** or **size** of a shape.

◎ A **reflection** is a transformation involving a **mirror line** (or **axis of symmetry**), in which the line from each point to its image is perpendicular to the mirror line and has its midpoint on the mirror line.

To describe a reflection fully, you must describe the position or give the equation of its mirror line.

Triangle A has been reflected in the mirror line $y = 1$ to give the image B.

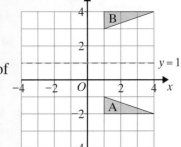

◎ A **rotation** is a transformation in which the shape is turned about a fixed point, the **centre of rotation**. To describe a rotation fully, you must give the **centre**, **angle** and **direction**.

A positive angle is anticlockwise and a negative angle is clockwise.

Triangle A is rotated about the origin through 90° anticlockwise to give the image C.

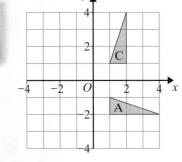

◎ A **translation** is a transformation where every point of the **object** moves the same distance and direction to map onto the **image**.

The top number is the movement across (positive to the right, negative to the left).

The bottom number is the vertical movement (positive upwards, negative downwards).

Triangle A has been mapped onto triangle B by a translation of 3 units to the right and 2 units down. The **translation vector** to map A to B is $\begin{pmatrix} 3 \\ -2 \end{pmatrix}$.

To return triangle B to triangle A use a translation of 3 units to the left and 2 units up. The **translation vector** to map B to A is $\begin{pmatrix} -3 \\ 2 \end{pmatrix}$.

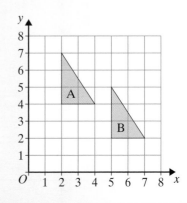

◉ An **enlargement** is a transformation that changes the size of an object but not its shape. To describe an enlargement fully you must give the **centre of enlargement** and the **scale factor**.

> The object and the image are <u>similar</u> because they are the same shape but a different size. Each side of triangle B is a third of the corresponding side of triangle A.

> A scale factor equal to 1 doesn't change the shape.
> A scale factor greater than 1 enlarges the shape.
> A scale factor between 0 and 1 shrinks the shape.
> A negative scale factor inverts the shape.

Triangle B is an enlargement of triangle A.
The **centre of enlargement** is (1, 7).

The **scale factor** is $\frac{1}{3}$.

> Scale factor = $\dfrac{\text{length of line on enlargement}}{\text{length of line on original}}$

◉ **Dimensions**

Length is measured in units (one-dimensional)
Area is measured in square units (two-dimensional), for example cm^2, m^2
Volume is measured in cubic units (three-dimensional), for example cm^3, m^3

Vectors

> To get from A to B go right 3 units and down 2 units.

◉ **Vectors** have direction and magnitude.
Scalars have magnitude only.

> Magnitude means size so scalars are just numbers.

$$\overrightarrow{AB} = \mathbf{a} = \begin{pmatrix} 3 \\ -2 \end{pmatrix}, \quad \overrightarrow{BC} = \mathbf{b} = \begin{pmatrix} 3 \\ 3 \end{pmatrix}$$

$$\overrightarrow{AC} = \overrightarrow{AB} + \overrightarrow{BC}$$

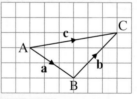

> Trace your finger along a then b in the direction of the arrows. You are starting at A and ending at C which gives the same result as tracing your finger along c.

$$\mathbf{c} = \mathbf{a} + \mathbf{b} = \begin{pmatrix} 3 \\ -2 \end{pmatrix} + \begin{pmatrix} 3 \\ 3 \end{pmatrix} = \begin{pmatrix} 6 \\ 1 \end{pmatrix}$$

$$\overrightarrow{AB} = \overrightarrow{AC} + \overrightarrow{CB} = \overrightarrow{AC} - \overrightarrow{BC}$$

$$\mathbf{a} = \mathbf{c} - \mathbf{b} = \begin{pmatrix} 6 \\ 1 \end{pmatrix} - \begin{pmatrix} 3 \\ 3 \end{pmatrix} = \begin{pmatrix} 3 \\ -2 \end{pmatrix}$$

> Going from C to B is the reverse of following b, so $\overrightarrow{CB} = -\overrightarrow{BC} = -\mathbf{b}$.

◉ If $\mathbf{a} = \begin{pmatrix} x \\ y \end{pmatrix}$ and $\mathbf{b} = m\begin{pmatrix} x \\ y \end{pmatrix}$, then $\mathbf{a}$ and $\mathbf{b}$ are parallel and $\mathbf{b}$ is m times the length of $\mathbf{a}$.

If $m = -1$, then $\mathbf{a}$ and $\mathbf{b}$ are equal, parallel and opposite.

If $\mathbf{a} = \begin{pmatrix} x \\ y \end{pmatrix}$ and $\mathbf{b} = \begin{pmatrix} -y \\ x \end{pmatrix}$, then $\mathbf{a}$ and $\mathbf{b}$ are perpendicular.

Worked Example Combining transformations

Question

The diagram shows a shaded triangle.

a Reflect the triangle in the line $y = x$.
Label the image A.

AQA
EXAMINER SAYS...
Make sure you label the images so the examiner can give full marks.

b Rotate triangle **A** 90° clockwise about the point $(-1, 2)$.
Label the image B.

c Reflect triangle **B** in the line $y = 0$
Label the image C.

d Describe fully the **single** transformation that maps the original shaded triangle onto triangle **C**.

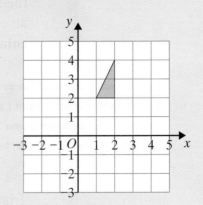

Solution

a

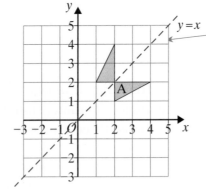

Draw in the mirror line first and use tracing paper to help.

GET IT RIGHT!

Be extra careful when reflecting in $y = x$. Turn your page so that the mirror line is vertical as this will help you draw the image.

GET IT RIGHT!

Remember that $y = 0$ is the horizontal x-axis.

b

c

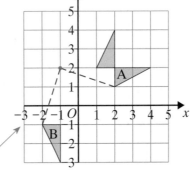

GET IT RIGHT!

Remember to use the given centre of rotation and not the origin.

Use tracing paper to help. Put your pencil point on $(-1, 2)$ and turn the paper through a quarter turn.

d The original shape is translated 3 units left and 1 unit down, so the transformation is a translation of $\begin{pmatrix} -3 \\ -1 \end{pmatrix}$.

Remember the top number is the movement across – it is negative here to show it is to the left.

Remember the bottom number is the vertical movement – it is negative here to show it is downwards.

BUMP UP
A
THE GRADE

To get a grade **A** you must be able to enlarge a shape using a negative scale factor so make sure you understand the following type of question.

Worked Example Enlargement

(Question) In the diagram, shape B is an enlargement of shape A.

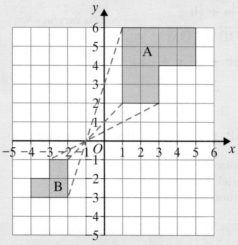

GET IT RIGHT!

When an enlargement reduces the size of the shape the scale factor will be a fraction between 0 and 1.
Here it will also be negative because the shape is the other side of the centre of enlargement.

Join corresponding vertices on the diagram. Where the rays cross is the centre of enlargement. The rays cross <u>between</u> the shapes which means the scale factor will be <u>negative</u>.

a Write down the coordinates of the centre of enlargement.

b Write down the scale factor of enlargement.

EXAMINER SAYS...

The scale factor of the enlargement should not be given as a ratio and you shouldn't put any units.

(Solution)
a The centre of enlargement is $(-1, 0)$

b The scale factor is $-\dfrac{1}{2}$

Be careful when writing your coordinates. The answer is **not** $(0, -1)$ or $(1, 0)$. Don't forget the brackets.

BUMP UP

A/A*

THE GRADE To get a grade **A** you need to be able to answer questions involving vectors. For a grade **A*** the vector problems will be based on more complex vector geometry as in the next example.

Worked Example **Vectors**

(Question)
OAB is a triangle.

$\overrightarrow{OA} = 3\mathbf{a}$ and $\overrightarrow{OB} = 3\mathbf{b}$.

M is the midpoint of AB.

N is the midpoint of OA.

P is the point on OM such that OP : PM = 2 : 1

a Find, in terms of **a** and **b**, the vectors

   i $\overrightarrow{BN}$ **ii** $\overrightarrow{OM}$ **iii** $\overrightarrow{BP}$

b Show that B, P and N are collinear.

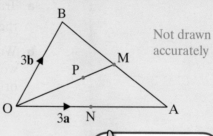

Not drawn accurately

GET IT RIGHT!

The ratio means that OP = $\frac{2}{3}$ OM

GET IT RIGHT!

$\overrightarrow{BO} \neq \overrightarrow{OB}$
Direction matters!

(Solution)
a i $\overrightarrow{BN} = \overrightarrow{BO} + \overrightarrow{ON} = -\overrightarrow{OB} + \frac{1}{2}\overrightarrow{OA}$

$\overrightarrow{BO} = -\overrightarrow{OB}$ because it is the same length in the opposite direction.

$= -3\mathbf{b} + \frac{1}{2} \times 3\mathbf{a}$

$\overrightarrow{ON} = \frac{1}{2}\overrightarrow{OA}$ because N is the midpoint of OA.

$= \frac{3}{2}\mathbf{a} - 3\mathbf{b}$

ii $\overrightarrow{OM} = \overrightarrow{OB} + \overrightarrow{BM} = \overrightarrow{OB} + \frac{1}{2}\overrightarrow{BA}$

$= \overrightarrow{OB} + \frac{1}{2}(\overrightarrow{BO} + \overrightarrow{OA})$

$= 3\mathbf{b} + \frac{1}{2}(-3\mathbf{b} + 3\mathbf{a})$

$= \frac{3}{2}\mathbf{a} + \frac{3}{2}\mathbf{b}$

$\overrightarrow{BA}$ is the same result as going from B to O, then O to A, so

$\overrightarrow{BA} = \overrightarrow{BO} + \overrightarrow{OA}$

iii $\overrightarrow{BP} = \overrightarrow{BO} + \overrightarrow{OP} = -\overrightarrow{OB} + \frac{2}{3}\overrightarrow{OM}$

$= -3\mathbf{b} + \frac{2}{3}\left(\frac{3}{2}\mathbf{a} + \frac{3}{2}\mathbf{b}\right)$

$= \mathbf{a} - 2\mathbf{b}$

Be careful when multiplying out brackets and collecting terms.

b $\overrightarrow{BN} = \frac{3}{2}\mathbf{a} - 3\mathbf{b} = \frac{3}{2}(\mathbf{a} - 2\mathbf{b}) = \frac{3}{2}\overrightarrow{BP}$

As $\overrightarrow{BN}$ is a multiple of $\overrightarrow{BP}$, the points B, P and N are collinear.

Because $\overrightarrow{BN}$ is a multiple of $\overrightarrow{BP}$ they are parallel. They also have the point B in common so B, P and N lie on the same line.

Transformations

1 The diagram shows two triangles A and B.

  a Describe fully the single transformation that maps triangle A onto triangle B.

  b Copy the diagram and draw the image of triangle A after it is reflected in the line $y = x$.

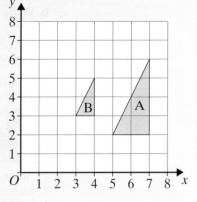

2 Copy the axes and shaded shape X.

  a Translate the shaded shape X by the vector $\begin{pmatrix} 3 \\ -1 \end{pmatrix}$.

  b Write down the translation vector that would return the image of X back to its original position.

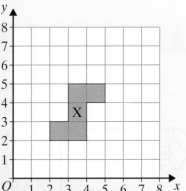

3 The length, width and height of this cuboid are all different.

Planes cut the cuboid into two equal parts as shown in the diagram.

Which of these are planes of symmetry?

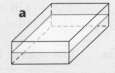 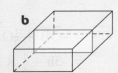 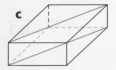 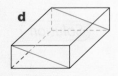

4 The diagram shows a shaded triangle.

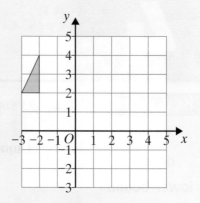

a Reflect the triangle in the line $x = 1$.

Label the image A.

b Reflect your image, triangle A, in the line $y = x$.

Label the image B.

c Describe fully the **single** transformation that maps the original triangle onto triangle B.

5 In each of the following expressions p, q and r represent lengths.

State whether each expression could represent a length, an area, a volume or none of these.

a pr **b** $2(p + q + r)$ **c** $\dfrac{pq}{r}$ **d** $p(q^2 + r^2)$ **e** $\dfrac{p + q}{r}$

6 Copy the diagram and enlarge the shape with scale factor −2.

Use the point (3, 5) as the centre of enlargement.

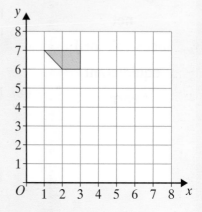

7 In the diagram $\overrightarrow{AB} = 2\mathbf{a}$, $\overrightarrow{EB} = 2\mathbf{b}$ and $\overrightarrow{AE} = \overrightarrow{EC}$.

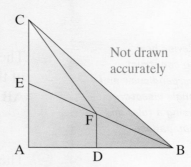

Not drawn accurately

a Find, in terms of $\mathbf{a}$ and $\mathbf{b}$, the vectors

i $\overrightarrow{AE}$

ii $\overrightarrow{AC}$

b D and F are the midpoints of AB and EB.

i Find $\overrightarrow{DF}$ in terms of $\mathbf{a}$ and $\mathbf{b}$.

ii Show that DF is parallel to AE.

Measures and drawing

Key points

◎ **Compound measures**

Cover up the quantity you want and then follow the signs.

$$\text{Speed} = \frac{\text{distance}}{\text{time}}$$
$$\text{Distance} = \text{speed} \times \text{time}$$
$$\text{Time} = \frac{\text{distance}}{\text{speed}}$$

$$\text{Density} = \frac{\text{mass}}{\text{volume}}$$
$$\text{Mass} = \text{density} \times \text{volume}$$
$$\text{Volume} = \frac{\text{mass}}{\text{density}}$$

◎ **Minimum and maximum values** – the **lower bound** is the minimum possible value and the **upper bound** is the maximum possible value of a measurement.

If a length is 5 cm correct to the nearest centimetre, the lower bound = 4.5 cm and the upper bound = 5.5 cm

◎ **Plans and elevations**

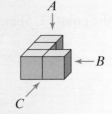

Plan view (from A)   **Side elevation** (B)   **Front elevation** (C)

Revise how to construct perpendicular and angle bisectors using a ruler and compasses.

◎ The **perpendicular bisector**, CD, of the line AB is at right angles to AB and divides it in half.

CD is the perpendicular bisector of AB, so AX = XB and angle CXB = 90°

◎ The **angle bisector**, AZ, of the angle BAC divides the angle in half.

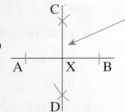

AZ is the angle bisector of angle BAC, so angle BAZ = angle ZAC.

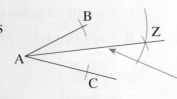

⌾ Constructing angles

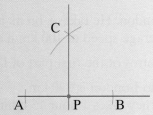

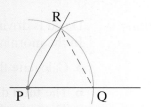

To construct an angle of 90°:

1 Point of compasses on P, draw arcs at A and B.

2 Point on A, then B, draw arcs that cross at C.

3 Join CP.

To construct an angle of 60°:

1 Point of compasses on P, draw a large arc.

2 Point on Q, draw arc with same radius to cross first arc at R.

3 Join PR.

⌾ The **locus** is the path of a moving point, for example:

The locus of points equidistant from A and B is the perpendicular bisector of AB.

> *The plural of locus is loci.*

> *Equidistant means the same distance.*

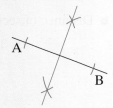

The locus of points equidistant from AB and AC is the angle bisector of angle BAC.

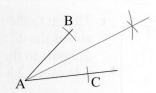

The locus of points a fixed distance from O is a circle. The equation of a circle, centre the origin and radius r is

$$x^2 + y^2 = r^2$$

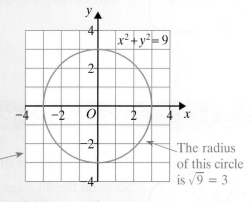

> In this circle, for every point on the circumference the sum of the squares of the x- and y-coordinates is 9. For example, the point (3, 0) is on the circle because $3^2 + 0^2 = 9$.

The radius of this circle is $\sqrt{9} = 3$

A/A*

THE GRADE To get a grade **A/A*** you must be able to find the upper and lower bounds of more difficult calculations with quantities given to required degrees of accuracy.

Worked Example Speed and upper and lower bounds

Question

David is driving from Leeds to London. He takes a break at lunchtime.
In the morning he drives at an average speed of 100 km/h and takes 1 hour 45 minutes.

a Calculate the distance in kilometres of the first part of David's journey.

b The **total** distance from Leeds to London is 320 km.
The second part of David's journey takes 1 hour 15 minutes.
What is his average speed for the second part of his journey?

c David returns home by train.
The train journey takes $2\frac{1}{2}$ hours, to the nearest 5 minutes.
The distance of 320 km is given to the nearest 10 kilometres.
What is the lower bound of the average speed of the train?

Solution

a Time $= 1$ hour 45 min $= 1\frac{3}{4}$ h $= 1.75$ h

distance $=$ speed $\times$ time $= 100 \times 1.75 = 175$ km

Be careful!
1 hour 45 min is
not 1.45 hours.

b Distance of second part of journey $= 320 - 175 = 145$ km

Time of second part $= 1$ hour 15 min $= 1\frac{1}{4}$ h $= 1.25$ h

$$\text{speed} = \frac{\text{distance}}{\text{time}}$$

$$\text{speed} = \frac{145}{1.25} = 116 \text{ km/h}$$

The upper bound
of the time is **not**
2 h 34 min or 2 h
34.9 min.

c 320 km to the nearest 10 km
means the distance is between 315 km and 325 km.
$2\frac{1}{2}$ hours to the nearest 5 minutes means the time
is between 2 h 25 min and 2 hr 35 min.

$$\text{Lower bound of speed} = \frac{\text{minimum distance}}{\text{maximum time}}$$

$$= \frac{315}{2\text{h } 35 \text{ min}} = \frac{315}{2.583...}$$

$$= 121.935... \text{ km/h} = 121.94 \text{ km/h (2 d.p.)}$$

Work out the
individual lower
and upper bounds
and remember
that minimum
$(x \div y) =$ **minimum**
$x \div$ **maximum** y.

Worked Example Loci and construction

Question

The diagram shows three mobile phone masts A, B and C.

Show on the diagram the region that is less than 6 km from masts A and B and is closer to mast B than mast C

A ×

Scale: 1 cm represents 2 km

× B

× C

Solution

Draw two large arcs, radius 3 cm, centre A and B. The region enclosed by the arcs is less than 6 km from A and B

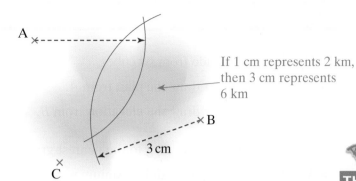

If 1 cm represents 2 km, then 3 cm represents 6 km

Construct the perpendicular bisector of CB. To the right of the bisector is the region of all points nearer B than C. Shade your answer clearly.

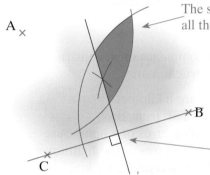

The shaded area satisfies all the constraints

The perpendicular bisector of BC is the locus of all points that are the same distance from masts B and C

BUMP UP

A/A*

THE GRADE To get a grade **A** you need to be able to construct the graphs of loci such as circles. For an **A*** you also need to be able to solve difficult questions involving loci and graphs. Look at the following example.

Worked Example ## Graphs of loci

Question

a Draw the graph of $x^2 + y^2 = 4$

b Use your graph to solve the simultaneous equations

$$y = 2x - 2$$
$$x^2 + y^2 = 4$$

AQA

EXAMINER SAYS...

The question tells you to use your graph, so don't try to solve the equations using a different method.

Solution

a $x^2 + y^2 = 4$ is the equation of a circle, centre the origin.

$r^2 = 4$, so $r = 2$

GET IT RIGHT!

The equation of a circle is $x^2 + y^2 = r^2$ so here the radius is 2, **not** 4.

The points of intersection of the circle and $y = 2x - 2$ are the solutions of the simultaneous equations.

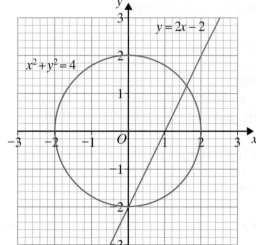

The line $y = 2x - 2$ has a gradient of 2 and an intercept on the y-axis of -2

b The solutions are $x = 0$, $y = -2$ and $x = 1.6$, $y = 1.2$

Measures and drawing

Can you complete these questions in **40** minutes?

1 This 3-D solid is made from six cubes.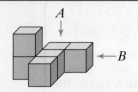

   a On squared paper draw the plan from *A*.

   b On squared paper draw the elevation from *B*.

2 A packet of crisps weighs 30 g to the nearest gram.

   a What are the minimum and maximum weights of the packet?

   b The crisps are sold in economy packs of 6.
What are the minimum and maximum weights of an economy pack?

3 a Using a ruler and compasses only, construct an angle of 60°.
Show all your construction arcs.

   b Two ships A and B are 6 kilometres apart.
A is on a bearing of 270° from B.
Ship A is within 4 kilometres of a lighthouse.
Ship B is within 5 kilometres of the lighthouse.
Using a scale of 1 cm to represent 1 km,
accurately construct a diagram showing the position of the ships.
Shade the possible area in which the lighthouse could be.

4 Draw an 8 cm line.
Draw the locus of all points 2 cm from the line.

5 Ben and Zoë are travelling from North Green to West Park.
Ben rollerblades 6 miles along a cycle path at an average speed of 8 mph.
Zoë walks the 4 miles directly at an average speed of 3 mph.
Calculate the difference in time between the two journeys.
Give your answer in minutes.

6 A greyhound runs 100 metres in 8.95 seconds.
Estimate its average speed in kilometres per hour.
You **must** show your working.

7 ABC is a triangle.
AB = 6 cm, AC = 4 cm and angle CAB = 90°.
Construct triangle ABC using a ruler and compasses only.
Show your construction arcs.
Measure and write down the length of BC.

8 A cube has a volume of 100 cm³, measured to the nearest cubic centimetre.
What is the lower bound of the length of a side of the cube?

9 Find the equation of the circle, centre (0, 0), passing through the point (3, 4).

Shape, space and measures

5

Pythagoras' theorem

Key words

hypotenuse

Pythagoras' theorem

Pythagorean triple

similar

congruent

Key points

◎ **Pythagoras' theorem** – in a **right-angled triangle**, the square of the length of the **hypotenuse** is equal to the sum of the squares of the lengths of the other two sides.

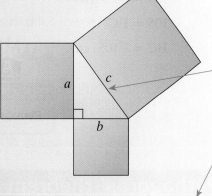

$$c^2 = a^2 + b^2$$

> The hypotenuse is the longest side and is opposite the right angle.

> For example, if $c = 17$ cm, $a = 8$ cm and $b = 15$ cm, then $c^2 = a^2 + b^2$ so this is a right angle.

◎ **Converse of Pythagoras' theorem** – in any triangle if $c^2 = a^2 + b^2$ then the triangle has a **right angle** opposite side c.

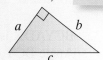

◎ A **Pythagorean triple** is a set of three integers a, b, c that satisfies $c^2 = a^2 + b^2$

It helps to learn common Pythagorean triples as they are often used on the non-calculator paper. For example,

3, 4, 5 ($5^2 = 3^2 + 4^2$),

5, 12, 13 ($13^2 = 5^2 + 12^2$)

◎ Shapes are **similar** if they are equiangular and the ratio of the corresponding sides is the same, that is, one is an enlargement of the other. Their areas and volumes will be in related ratios.

> In an enlargement the angles stay the same.

◎ If the **linear** ratio of a solid is $a:b$,
then the **area** ratio will be $a^2:b^2$
and the **volume** ratio will be $a^3:b^3$

◎ To prove that two triangles are **congruent** (have the same size and shape) you need to show one of the following:

– 3 corresponding sides are equal (SSS)
– 2 angles and 1 corresponding side are equal (AAS)
– 2 corresponding sides and the angle between those sides are equal (SAS),
– both triangles have a right angle, hypotenuse and another side equal (RHS).

> Remember that matching angles (AAA) only shows the triangles are <u>similar</u> not congruent.

> The angle must be <u>between</u> the sides to prove congruence.

Worked Example Pythagoras' theorem

Question

ABC is a right-angled triangle.
AB = 8 cm and AC = 13 cm.

Calculate the length of BC.
Leave your answer as a square root.

Solution

Using Pythagoras' theorem in triangle ABC,

$$AC^2 = AB^2 + BC^2$$ Put AC on the
$$13^2 = 8^2 + BC^2$$ left as it is the hypotenuse
$$169 = 64 + BC^2$$
$$105 = BC^2$$ ←— Subtract 64 from both sides
$$BC = \sqrt{105} \text{ cm}$$

GET IT RIGHT!

This question could be set on the non-calculator paper since you are not asked to find the square root of 105.

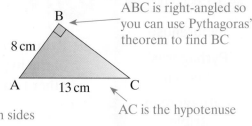

ABC is right-angled so you can use Pythagoras' theorem to find BC

8 cm

A 13 cm C

AC is the hypotenuse

BUMP UP

A

THE GRADE To get a grade **A** you must be able to use Pythagoras' theorem in 3-D problems.

Worked Example Pythagoras' theorem in 3-D

Question

BCYADX is a triangular prism.
Angle BCY = 90°
AB = 12 cm, BC = 4 cm, CY = 3 cm

Calculate the length of AY.

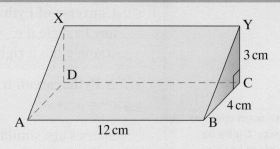

X   Y
                     3 cm
    D
              C
            4 cm
A   B
      12 cm

Solution

AY is the hypotenuse of triangle ABY.
To use Pythagoras' theorem in this triangle you need to work out BY.

Using Pythagoras' theorem in triangle BCY,

$$BY^2 = BC^2 + CY^2$$
$$BY^2 = 4^2 + 3^2 = 16 + 9 = 25$$
$$BY = \sqrt{25} = 5 \text{ cm}$$

Now in triangle ABY,

$$AY^2 = AB^2 + BY^2$$
$$AY^2 = 12^2 + 5^2 = 144 + 25 = 169$$
$$AY = \sqrt{169} = 13 \text{ cm}$$

GET IT RIGHT!

If you spot the 3, 4, 5 and 5, 12, 13 triples you can just quote them and save yourself all the working out.

Y

A 12 cm B

Y
3 cm
B 4 cm C

Draw a sketch to see which sides you need to work out.

Y
5 cm
A 12 cm B

THE GRADE To get a grade **A** you need to be able to work with volumes and areas of similar shapes. The following question shows you how to calculate a volume from a linear ratio.

Worked Example Similar shapes

(Question)

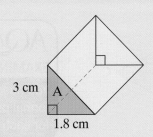

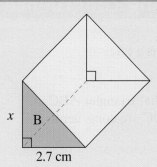

The diagram shows two similar triangular prisms, A and B.

a Find x.

b The volume of prism B is 54 cm³.
Calculate the volume of prism A.

(Solution)

Don't approximate $\frac{2}{3}$ as a decimal as this could lead to rounding errors.

a Since the triangles are similar the corresponding sides are in the same ratio,

$1.8 : 2.7 = 18 : 27 = 2 : 3$ ◄—— Simplify the ratio

$$\frac{3}{x} = \frac{2}{3}$$

$$\frac{x}{3} = \frac{3}{2}$$ ◄—— Turn both sides upside down so that x is on the top

$$3 \times \frac{x}{3} = \frac{3}{2} \times 3$$ ◄—— Multiply both sides by 3

$$x = \frac{3}{2} \times 3$$

$$x = 4.5 \text{ cm}$$

The ratio of the volumes is always the cube of the ratio of the lengths.

b Linear ratio of A and B is $2 : 3$

Volume ratio is $2^3 : 3^3 = 8 : 27$ ◄ Cube both sides of the ratio

$$\frac{\text{volume of A}}{\text{volume B}} = \frac{8}{27}$$

$$\frac{\text{volume of A}}{54} = \frac{8}{27}$$

$$54 \times \frac{\text{volume of A}}{54} = \frac{8}{27} \times 54$$ ◄—— Multiply both sides by 54

$$\text{volume of A} = \frac{8}{27} \times 54$$

$$\text{volume of A} = 16 \text{ cm}^3$$

THE GRADE To get a grade **A** you must be able to prove that two triangles are congruent. Practise by proving the construction theorems that you used in the last chapter.

Worked Example Proving congruence

(Question) ABCD is a kite.

Prove that triangles ABC and ADC are congruent.

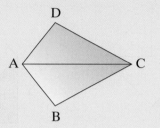

AQA
EXAMINER SAYS...
You need to present a formal proof. Give a reason for each statement.

(Solution)

This symbol means 'is congruent to'. Remember to write the corresponding vertices in the same order.

In triangle ABC and triangle ADC,

AB = AD (adjacent sides of a kite)

BC = CD (adjacent sides of a kite)

AC = AC (same line)

So triangle ABC ≡ triangle ADC (SSS)

You could also have used angle ABC = angle ADC (opposite angles of a kite) for your third reason. This would have been using the SAS proof.

Pythagoras' theorem

END OF CHAPTER QUESTIONS

Time Yourself!

Can you complete these questions in **30** minutes?

1 DEF is a right-angled triangle.

DE = 12 cm and EF = 13 cm.

Calculate the length of the side DF.

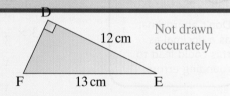

Not drawn accurately

2 Triangles ABC and DEF are similar.

a Calculate the length of BC.

b Calculate the length of DF.

Angle EFD = 19°.

c Write down the size of angle BCA.

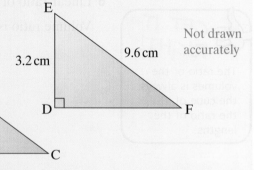

Not drawn accurately

3 ABCD is a quadrilateral.

Angles ABD and BDC are right angles.

AB = 12 cm, AD = 15 cm and DC = 7 cm.

Show that the length of BC is $\sqrt{130}$ cm.

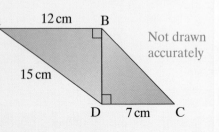

Not drawn accurately

4 In the diagram, angle CAB = angle CED

BC = DC = 6.3 cm
AC = 2.6 cm and AB = 9.4 cm.

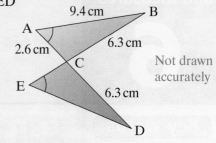

Not drawn accurately

a Prove that the two triangles, ABC and EDC, are congruent.

b State the length of EC.

5 PQRSTUVW is a cube of side 2 cm.

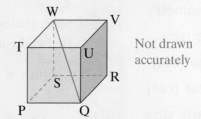

Not drawn accurately

Calculate the length of WQ.

Leave your answer in surd form.

6 A supermarket sells two similar cartons of detergent.

The smaller pack is 25 cm high.

The larger pack shown in the diagram is 36 cm high.

Is the claim on the larger pack justified?

You **must** show your working.

Not drawn accurately

6

Trigonometry

In the diagram, angle *CAB* = angle *CED*

BC = *DC* = 4.2 cm

AC = 2.6 cm and *AB* = 3.1 cm

4   Prove that the two triangles *ABC* and *DEC* are congruent.

Find the missing length...

Key words

trigonometry

sine (sin)

cosine (cos)

tangent (tan)

opposite side

adjacent side

hypotenuse

Key points

Sine, cosine and tangent

In a **right-angled** triangle,

$$\sin x = \frac{opposite}{hypotenuse}$$

$$\cos x = \frac{adjacent}{hypotenuse}$$

$$\tan x = \frac{opposite}{adjacent}$$

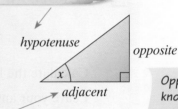

The longest side.

hypotenuse

opposite

Opposite the angle you know or want to find.

Next to the angle you know or want to find.

Trigonometric graphs

The values of sine and cosine repeat every 360°.

The values of tangent repeat every 180°.

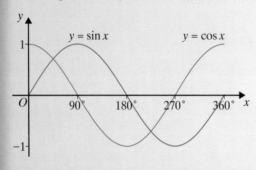

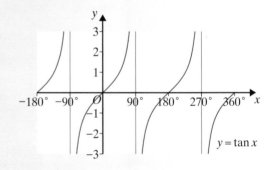

Sine rule

The sine rule says:

$$\frac{\sin A}{a} = \frac{\sin B}{b} = \frac{\sin C}{c}$$

or $$\frac{a}{\sin A} = \frac{b}{\sin B} = \frac{c}{\sin C}$$

Label the side a opposite angle A etc.

Cosine rule

The cosine rule says:

$$a^2 = b^2 + c^2 - 2bc\cos A$$

or $$b^2 = a^2 + c^2 - 2ac\cos B$$

or $$c^2 = a^2 + b^2 - 2ab\cos C$$

Area of a triangle

The area of any triangle can be found by

$$\text{Area} = \tfrac{1}{2}\,ab\sin C = \tfrac{1}{2}\,ac\sin B$$

$$= \tfrac{1}{2}\,bc\sin A$$

Worked Example · Trigonometry

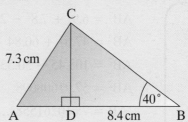

(Question)

ABC is a triangle.
CD is the perpendicular
from the point C to the line AB.
AC = 7.3 cm, DB = 8.4 cm
and angle CBD = 40°.

a Calculate the length CD.

b Calculate the angle CAD.

(Solution)

a
$$\tan 40° = \frac{opp}{adj}$$

$$\tan 40° = \frac{CD}{8.4}$$

$$8.4 \times \tan 40° = CD$$

$$CD = 8.4 \times 0.8390996312\ldots = 7.048436902\ldots$$

$$CD = 7.05 \text{ cm (3 s.f.)}$$

Draw and label
a sketch of the
triangle that you
are working with.

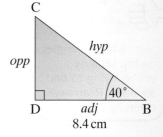

GET IT RIGHT!

Choose the
right ratio. You
want the side
opposite the angle
and know the
adjacent side, so
use tan.

b
$$\sin x = \frac{opp}{hyp}$$

$$\sin x = \frac{7.048436902\ldots}{7.3}$$

$$\sin x = 0.9655393016\ldots$$

$$x = \sin^{-1} 0.9655393016\ldots$$

$$x = 74.91\ldots = 74.9° \text{ (1 d.p.)}$$

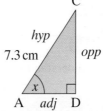

The inverse of sine
is written as sin⁻¹
or arcsin.

AQA

EXAMINER SAYS…

Use the full value
from part **a** to
avoid rounding
errors.

GET IT RIGHT!

You know the
opposite and the
hypotenuse so use
sin.

A/A*

THE GRADE To get a grade **A/A*** you must be able to use
the sine and cosine rules confidently.

Worked Example · Sine and cosine rules

(Question)

ABC is a triangle.

AC = 7.8 cm, BC = 6.9 cm

and angle ACB = 62°.

a Calculate AB.

b Calculate angle ABC.

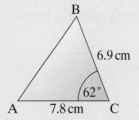

ABC is not a right-
angled triangle so
you need to use the
sine and cosine rules

Solution

GET IT RIGHT!

Use the rule correctly. Only the 2*ab* part is multiplied by cos C, not the whole right-hand side.

GET IT RIGHT!

Start with the angles on top because you're trying to find an angle. Rearrange carefully.

a Using the cosine rule in triangle ABC,

$$c^2 = a^2 + b^2 - 2ab\cos C$$

$$AB^2 = 6.9^2 + 7.8^2 - 2 \times 6.9 \times 7.8 \times \cos 62°$$

$$AB^2 = 47.61 + 60.84 - 107.64\cos 62°$$

$$AB^2 = 108.45 - 50.53391902...$$

$$AB^2 = 57.91608098...$$

$$AB = 7.610261558... = 7.6 \text{ cm (1 d.p.)}$$

b Using the sine rule in triangle ABC,

$$\frac{\sin B}{b} = \frac{\sin C}{c}$$

$$\frac{\sin B}{7.8} = \frac{\sin 62°}{7.610261558}$$

$$\sin B = \frac{7.8 \times \sin 62°}{7.610261558}$$

$$\sin B = 0.9049611727...$$

angle ABC = $\sin^{-1} 0.9049611727...$

$$= 64.81... = 64.8° \text{ (1 d.p.)}$$

Use the full value from part **a**

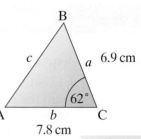

Draw and label a sketch of the triangle.

EXAMINER SAYS...

Make sure your calculator is in DEG mode before you start your calculation.

BUMP UP **A*** **THE GRADE** To get a grade **A*** you must be able to solve trigonometry problems in 3-D and understand graphs of trigonometric functions for angles of any size.

Worked Example **Graphs of trigonometric functions**

Question

AQA

EXAMINER SAYS...

This type of question would probably appear on the non-calculator paper.

The sketch shows the graph of $y = \tan x$ for $-180° \leqslant x \leqslant 180°$.

You are given that $\tan 30° = 0.577$

a Write down another solution of the equation $\tan x = 0.577$

b Solve the equation $\tan x = -0.577$ for $-180° \leqslant x \leqslant 180°$.

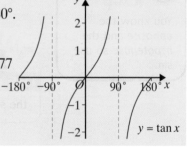

Solution

a From the graph,

$$x = -180° + 30° = -150°$$

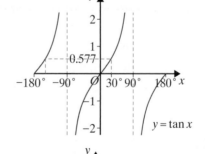

b From the graph, there are two solutions in the range,

$$x = -30° \text{ and}$$

$$x = 180° - 30° = 150°$$

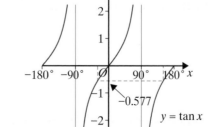

GET IT RIGHT!

Use the repeating pattern of the graph to help you find another angle. $x = 180° + 30° = 210°$ would have the same tangent but would not be correct here, because it's outside the range $-180° \leqslant x \leqslant 180°$

Trigonometry

1 a In triangle ABC, angle A is a right angle, angle C = 25° and BC = 9.4 cm.

Calculate the length of AC.

b In triangle DEF, angle E is a right angle, DE = 7 cm, EF = 4 cm.

Calculate the size of angle F.

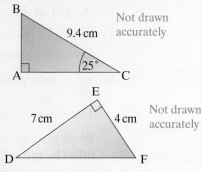

Not drawn accurately

2 The safety instructions say that a ladder must be inclined at between 72° and 78° to the ground.

a The diagram shows a ladder leaning against a wall. Is this ladder safe? You **must** show your working.

b A longer 6 metre ladder is to be used. Calculate the maximum vertical height that this ladder can safely reach.

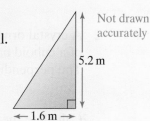

Not drawn accurately

3 a ABC is a triangle. AC = 10 cm, BC = 15 cm and angle BAC = 50°.

Calculate the size of angle ABC.

b DEF is a triangle. DE = 11 cm, DF = 12 cm and angle EDF = 60°.

Calculate the length of EF. Give your answer to an appropriate degree of accuracy.

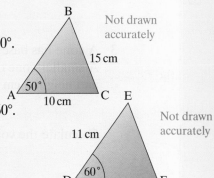

Not drawn accurately

4 You are given that $\sin 45° = \dfrac{\sqrt{2}}{2}$

Calculate the area of the triangle. Give your answer in simplified surd form.

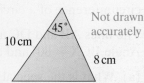

Not drawn accurately

5 The diagram shows a ramp into a building. The horizontal base, ABCD, and the sloping face, ABEF, are both rectangular.

Calculate the angle between the diagonal AE and the base.

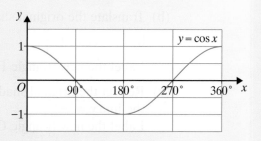

Not drawn accurately

6 The sketch shows the graph of $y = \cos x$ for $0° \leqslant x \leqslant 360°$. You are given that $\cos 75° = 0.2588$

a Write down another solution of the equation $\cos x = 0.2588$

b Solve the equation $\cos x = -0.2588$ for $0° \leqslant x \leqslant 360°$.

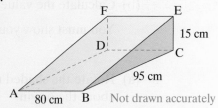

1   The diameter of a cylinder is 8 cm.
    The height of the cylinder is 6 cm.

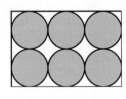

8 cm

6 cm

Not drawn
accurately

   (a) Work out the volume of the cylinder.
        Give your answer in terms of π.   *(3 marks)*

   (b) Six of the cylinders are packed in a crate of height 6 cm.
        The diagram shows the cylinders in the crate.
        The shaded area is the space between the cylinders.

        Work out the volume in the crate that is **not**
        filled by the cylinders.
        Give your answer in terms of π.   *(4 marks)*

2   A crystal ornament consists of a pyramid on top
    of a cuboid base as shown in the diagram.
    The perpendicular height of the pyramid is 6 cm.

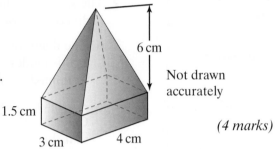

6 cm

Not drawn
accurately

1.5 cm

    Calculate the volume of the ornament.   *(4 marks)*

3 cm   4 cm

3   A cone has base radius 8 cm and height h cm.
    A smaller cone of base radius 4 cm
    and height 5 cm is cut from the top.

Not drawn
accurately

5 cm

h cm

4 cm

8 cm

    Calculate the volume of the frustum.   *(5 marks)*

4   A, B, C and D are points on the
    circumference of a circle.

    The lines AB and DC are produced to meet at X.
    Angle BCX = 80° and angle BXC = 25°

   (a) What is the special name for quadrilateral ABCD?   *(1 mark)*

   (b) Calculate the value of x.

        You **must** show your working.   *(3 marks)*

A   B   X
              25°

              80°
                  C
x
D

Not drawn
accurately

5   (a) Rotate the shaded triangle 90° anticlockwise
        about the origin.
        Label the new triangle A.   *(3 marks)*

   (b) Translate the original shaded triangle
        by the vector $\begin{pmatrix} 2 \\ -1 \end{pmatrix}$.
        Label the new triangle B.   *(1 mark)*

   (c) Reflect the original shaded triangle in
        the line $y = -1$
        Label the new triangle C.   *(2 marks)*

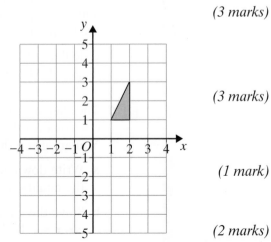

6 (a) Describe the transformation that maps triangle A to triangle B. *(2 marks)*

(b) Triangle A is an enlargement of triangle C.

   (i) Write down the scale factor of the enlargement. *(1 mark)*

   (ii) Write down the coordinates of the centre of the enlargement. *(1 mark)*

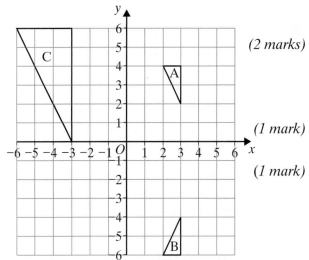

7 In the diagram $\overrightarrow{OR} = 3\mathbf{a}$, $\overrightarrow{RQ} = \mathbf{a}$, $\overrightarrow{OP} = 6\mathbf{b}$ and $\overrightarrow{PS} = 3\mathbf{b}$

M is the midpoint of QP.

(a) Find, in terms of **a** and **b**, simplifying your answers, the vectors:

   (i) $\overrightarrow{QP}$ (ii) $\overrightarrow{RM}$ *(3 marks)*

(b) Show clearly that points R, M and S lie on a straight line. *(3 marks)*

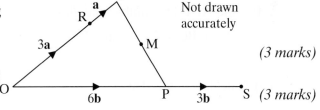

Not drawn accurately

8 Lee cycles in a 6 km race around a 240 m track.

(a) How many 240 m laps does Lee complete? *(3 marks)*

(b) On average Lee completes a 240 m lap in 30 seconds. Calculate Lee's average speed in metres per second. *(2 marks)*

(c) What is the total time that Lee takes to complete the 6 km race? Give your answer in minutes and seconds. *(3 marks)*

9 On the same axes, draw the graphs of $x^2 + y^2 = 16$ and $y = 3x - 1$ *(3 marks)*

Use your graphs to solve the simultaneous equations

$$y = 3x - 1$$

$$x^2 + y^2 = 16$$

(2 marks)

10 ABC is a right-angled triangle.

AC = 8 cm and AB = 5 cm.

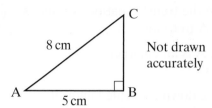

Calculate the length of BC.

Leave your answer as a square root.

(3 marks)

11 The diagram shows a flagpole that is 7 metres high. A wire support is attached to the top of the flagpole and fixed to the ground 2.5 metres from its base.

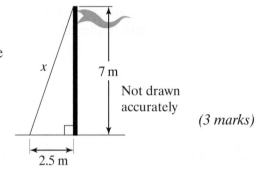

Calculate the length of the wire support, x.

(3 marks)

12 ABC and ADE are similar triangles.

AC = 3 cm, BC = 2.5 cm and DE = 15 cm.

(a) Calculate the length of CE.

(b) Calculate angle BAC.

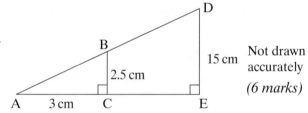

(6 marks)

13 ABCDE is a right pyramid on a square base. E is vertically above the centre of the square base.

EA = EB = EC = ED = 10 cm.

AB = 8 cm.

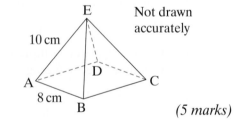

Calculate the angle between EA and the base ABCD.

(5 marks)

14 Amanda and Natalie work at the same office.

Amanda's home (A) is due east of the office (O).

Natalie's home (N) is north-west of the office.

The diagram shows their relative positions.

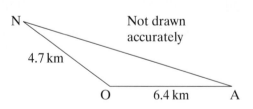

Calculate the distance, AN, from Amanda's home to Natalie's home.

(4 marks)

1   A concrete wrecking ball is used for demolishing walls.

It is in the shape of a cone on top of a hemisphere.

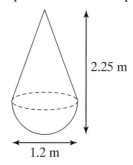

2.25 m

1.2 m

(a) Calculate the volume of the wrecking ball.

The candidate remembers to work with the radius. Using the diameter would mean a maximum of 2 marks for part **a.**

Substituting into the correct hemisphere formula attracts the first mark.

Radius of hemisphere = 1.2 ÷ 2 = 0.6 m

Volume of hemisphere = $\frac{1}{2} \times \frac{4}{3} \times \pi r^3 = \frac{1}{2} \times \frac{4}{3} \times \pi \times 0.6^3$

= 0.4523... m³

Don't round too early as this could lead to an accuracy error.

You are finding the volume of a hemisphere, so don't forget to halve it.

Height of cone = 2.25 − 0.6 = 1.65 m

A mark is awarded for finding the height of the cone.

A mark is awarded for at least one correct volume, either the hemisphere or the cone.

Volume of cone = $\frac{1}{3} \pi r^2 h = \frac{1}{3} \pi \times 0.6^2 \times 1.65$

= 0.6220... m³

Substituting into the correct cone formula attracts another mark

Total volume = 0.4523 + 0.6220 = 1.074... = 1.07 m³ (3 s.f.)

Volume = 1.07
Answer .... m³   *(5 marks)*

The last mark is awarded for the correct final answer.

(b) Calculate how many litres of concrete are used to make this wrecking ball.

1.07 m³ = 1.07 × 100 × 100 × 100 cm³

Remember 1 m = 100 cm, so 1 m³ = 100 × 100 × 100 cm³

= 1 070 000 cm³

Learn this fact before your exam.

The first mark is awarded for a correct method to convert into litres.

1000 cm³ = 1 litre, amount of concrete = $\frac{1\ 070\ 000}{1000}$ = 1070 litres

The final mark is awarded for the correct follow through answer. The candidate has converted the answer from part **a** into litres correctly.

1070
Answer .... litres   *(2 marks)*

2   (a) O is the centre of the circle.

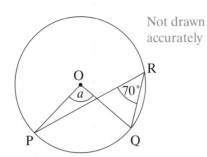

Not drawn
accurately

Be careful! Angles POQ
and PRQ are not equal
because they are not both
at the circumference.

Find the value of *a* giving a reason for your answer.

angle POQ = 2 × angle PRQ = 2 × 70° = 140°

The first mark is for
the correct value of *a*.

140

The second mark is awarded for giving
the correct geometrical reason.

Answer *a* = ... degrees

Reason ...the angle at the centre is twice the angle at the circumference.......

(2 marks)

(b) The points B and C lie on a circle centre O.

AB is a tangent to the circle at B.

Angle CAB = 50° and angle COB = 100°.

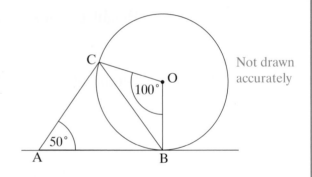

Not drawn
accurately

Prove that triangle CAB is isosceles.

Don't work
backwards by
assuming CAB
is isosceles
and showing
the angles
match. This is
not proving the
statement.

OB = OC (radii)

angle OBC = $\dfrac{180° - \text{angle BOC}}{2}$ (base angle of isosceles triangle)

$= \dfrac{180° - 100°}{2} = 40°$

A mark is awarded for finding
angle OBC and giving a reason.

angle OBA = 90° (angle between tangent and chord)

angle CBA = angle OBA - angle OBC

$= 90° - 40° = 50°$

A second mark is awarded for using angle
OBA = 90° (with a reason) to find angle CBA.

angle CAB = angle CBA = 50° so CAB is an isosceles triangle.

The final mark is for stating
why triangle CAB is isosceles

(3 marks)

Symbols, sequences and equations

Key words

sequence

*n*th term

product

simplify

expand

factorise

subject of a formula

integer

trial and improvement

Key points

- A **sequence** is a list of numbers that are connected in some way.
- The **nth term** is the general term in a sequence.
- The **product** of two numbers is the result of multiplying them together.
- To **simplify** means to collect like terms (adding or subtracting terms with the same variable).
- To **expand** means to remove brackets by multiplying.
- To **factorise** means to put brackets in an expression by finding common factors.
- The **subject of the formula** $A = \frac{bh}{2}$ is A, as it is a formula to work out A.
- An **integer** is a whole number, such as 4, 0 or –3
- **Trial and improvement** is a way of solving equations by making improved guesses.

Worked Example — Multiplying and dividing indices

Question

Simplify: **a** $3a^4b^2 \times 2a^3bc$ **b** $\frac{6a^5b^2}{3a^2b}$

Solution

a $3a^4b^2 \times 2a^3bc = 3 \times a^4 \times b^2 \times 2 \times a^3 \times b \times c$

$\qquad = 3 \times 2 \times a^4 \times a^3 \times b^2 \times b \times c$

$\qquad = 6\,a^7b^3\,c$ ←—Don't forget to multiply the 3×2 but just add the indices

> *b is the same as b^1*

b $\frac{6a^5b^2}{3a^2b} = 2a^3b$ ←—Divide 6 by 3, then subtract indices or cancel:

$$\frac{6 \times a \times a \times a \times a \times a \times b \times b}{3 \times a \times a \times b}$$

👍 **GET IT RIGHT!**

To **multiply** powers of a, just **add** the powers. To **divide** powers of a, just **subtract** the powers.

Worked Example — Expanding linear expressions

Question

Expand and simplify $(2y + 4)(3y - 2)$.

Solution

×	2y	+4
3y	6y²	+12y
−2	−4y	−8

$= 6y^2 + 12y - 4y - 8$

$= 6y^2 + 8y - 8$

> *Use the grid method.*

Worked Example Factorising

Question Factorise $4x^2 - 6xy$

Solution $4x^2 - 6xy$ $4x^2 = 2x \times 2x$
$= 2x(2x - 3y)$ $6xy = 2x \times 3y$

GET IT RIGHT!

Don't just take out 2 if you can take out $2x$.

Worked Example Factorising quadratic expressions

Question Factorise $2x^2 - 3x - 2$

The dark grey cells must multiply to the constant term (−2).

The light grey cells must add up to the linear term (−3x).

The only correct combination is shown below.

Solution

$\times$	$2x$	
x	$2x^2$	
		-2

$\times$	$2x$	$+1$
x	$2x^2$	$+x$
-2	$-4x$	-2

The solution is $2x^2 - 3x - 2 = (2x + 1)(x - 2)$

BUMP UP THE GRADE A/A*

To get a grade **A/A*** you must be able to recognise the difference of two squares, which have no x-term. For example, $x^2 - 16$ factorises as $(x + 4)(x - 4)$; the two terms in x are $-4x$ and $+4x$, which cancel each other out.

Worked Example Finding the nth term of a sequence

Question Find the nth term in this sequence:

5, 8, 11, 14, 17, ...

Solution nth term = difference $\times n$ + (first term − difference)

$= 3n + (5 - 3)$ Difference = 3
                                First term = 5

$= 3n + 2$

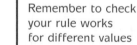

GET IT RIGHT!

Remember to check your rule works for different values of n.

Worked Example Equations with brackets

Question Solve the equation $3(2a + 4) = 2a + 4$

Solution $3(2a + 4) = 2a + 4$

$6a + 12 = 2a + 4$ ⟵ Multiply out the brackets first

$4a + 12 = 4$ ⟵ Subtract $2a$ from both sides

$4a = -8$ ⟵ Subtract 12 from both sides

$a = -2$ ⟵ Divide both sides by 4

AQA EXAMINER SAYS...

You should write the answer as $a = -2$, not just -2

Worked Example Equations with fractions

Question

Solve the equation $\frac{2x}{3} - 5 = 1$

Solution

$$\frac{2x}{3} - 5 = 1$$

$$\frac{2x}{3} = 6 \quad \longleftarrow \quad \text{Add 5 to both sides}$$

$$\frac{2x}{3} \times 3 = 6 \times 3 \quad \longleftarrow \quad \text{Multiply both sides by 3}$$

$$2x = 18$$

$$x = 9 \quad \longleftarrow \quad \text{Divide both sides by 2}$$

Undo the last operation first.

In this case, the last operation was –5

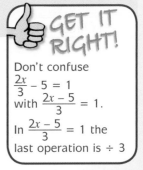

Don't confuse
$\frac{2x}{3} - 5 = 1$
with $\frac{2x - 5}{3} = 1$.
In $\frac{2x - 5}{3} = 1$ the last operation is ÷ 3

Worked Example Trial and improvement

Question

Use trial and improvement to find a positive solution to the equation

$x^2 - x = 18$. Give your answer to 2 decimal places.

Solution

EXAMINER SAYS...

Some candidates don't show every step of their working out. It is essential to do so in trial and improvement questions.

Trial value	$x^2 - x$	Comment
4	$16 - 4 = 12$	Too low
5	$25 - 5 = 20$	Too high
4.5	$20.25 - 4.5 = 15.75$	Too low
4.7	$22.09 - 4.7 = 17.39$	Too low
4.8	$23.04 - 4.8 = 18.24$	Too high
4.77	$22.7529 - 4.77 = 17.9829$	Too low
4.78	$22.8484 - 4.78 = 18.0684$	Too high
4.775	$22.800625 - 4.775 = 18.025625$	Too high

$x = 4.77$ (2 d.p.)

Once you know x is between 4.77 and 4.78, try halfway between them (4.775) to see whether the answer is closer to 4.77 or 4.78

Worked Example Changing the subject of a formula

Question

Rearrange the formula $A = \frac{(a + b)}{2}h$ to make a the subject.

Solution

$$A = \frac{(a + b)}{2}h \quad \longleftarrow \quad \text{Start with the formula}$$

$$\frac{A}{h} = \frac{(a + b)}{2}h \div h \quad \longleftarrow \quad \text{Divide both sides by } h$$

$$\frac{A}{h} = \frac{(a + b)}{2}$$

$$\frac{2A}{h} = a + b \quad \longleftarrow \quad \text{Multiply both sides by 2}$$

$$\frac{2A}{h} - b = a \quad \longleftarrow \quad \text{Subtract } b \text{ from both sides}$$

$$a = \frac{2A}{h} - b$$

The method is exactly the same as for solving an equation – use inverse operations.

The formula says:
Start with a; Add b; Divide by 2; Multiply by h.

The inverse operations used to undo this are:
Divide by h; Multiply by 2; Subtract b.

Worked Example — Changing the subject when the variable appears twice

Question   Rearrange the formula $a(x + b) = c(a - x)$ to make x the subject.

GET IT RIGHT!

Solution

$a(x + b) = c(a - x)$	Start with the formula
$ax + ab = ca - cx$	Expand
$ax + cx + ab = ac$	Add cx to both sides to put together both
$ax + cx = ac - ab$	Subtract ab from both sides
$x(a + c) = ac - ab$	Take out x as a common factor
$x = \dfrac{ac - ab}{a + c}$	Divide both sides by $a + c$

The method is similar to solving an equation – bring the terms containing x together. You then need to factorise so the new subject is written only once.

Symbols, sequences and equations

Time Yourself!

Can you complete these questions in **25** minutes?

1 Factorise $3x^2 + 6x^3$.

2 a Use the formula $v = u + at$ to find the value of v when $u = -6$, $a = 2.4$ and $t = 1.9$

  b Solve these equations.

   i $5x - 3 = 2 + 3x$ **ii** $3(2x + 1) = 7 - 5x$ **iii** $\dfrac{x + 7}{x + 4} = 3$

3 If $x = 2$ and $y = -5$, find the value of:

  a $3x + 2y$ **b** $\dfrac{x - 2y}{3}$

4 Use trial and improvement to find the positive solution to the equation $x^2 - x = 9$
  Give your answer to 1 decimal place.

5 Find the nth term of these sequences.

  a $1, 4, 9, 16, 25, \ldots$

  b $-3, 0, 5, 12, 21, \ldots$

6 Simplify:

  a $5a^2 \times 2a^3$ **b** $\dfrac{6p^3 \times 3p}{9p^2}$ **c** $3gh^4 \times 7g^2h$ **d** $\dfrac{18a^3b^5}{3a^4b^3}$

7 a Expand and simplify $4(5m - 3) + 2(6 - m)$

  b Factorise $n^2 - 49$

  c Make s the subject of the formula $w = s^2 - x$

  d Make x the subject of the formula $a(x - b) = a^2 + bx$

Algebra

2

Coordinates and graphs

Key words

Key words

coordinates

origin

axis (pl. axes)

gradient

intercept

midpoint

equidistant

speed

acceleration

Key points

- ◎ **Coordinates** identify a point.

- ◎ The **origin** is the point $(0, 0)$.

- ◎ The **axes** are the lines used to locate points.

- ◎ The **gradient** tells you is how steep a line is, calculated as
$$\frac{\text{change in vertical distance}}{\text{change in horizontal distance}}$$

- ◎ The **intercept** is the y-coordinate of the point where a line crosses the y-axis.

- ◎ The **midpoint** is the middle point of a line.

- ◎ If **A** and **B** are equidistant from **C**, the distance **AC** = the distance **BC**.

- ◎ A **conversion graph** is used to convert one unit into another, for example, pounds to kilograms.

- ◎ **Speed** is calculated as $\dfrac{\text{distance}}{\text{time}}$

- ◎ **Acceleration** is calculated as $\dfrac{\text{change of speed}}{\text{time}}$

(diagram: y-axis and x-axis with point P (3, 2); y-coordinate = 2, x-coordinate = 3, Origin at O)

Worked Example   Equations and straight lines

Question   Draw the line that has the equation, $y = 3x - 2$

Solution

GET IT RIGHT!

Don't forget that lines such as $x = 4$ are vertical (all the points on it have an x-coordinate of 4), and lines such as $y = -2$ are horizontal.

The graph of a straight line can always be written in the form $y = mx + c$, where m is the gradient, and c is the intercept on the y-axis.

So $y = 3x - 2$ crosses the y-axis at -2 and has a gradient of 3.

You can check your graph by making sure the coordinates fit the equation. For example, the graph passes through (2, 4), so when $x = 2$, $y = 4$:
When $x = 2$, $y = 3x - 2 = 3 \times 2 - 2 = 4$

Remember that equations can be written in other forms: $2x + y = 7$, $y = 7 - 2x$ and $y = -2x + 7$ are all the same equation. You should rearrange an equation into the form $y = mx + c$ (in this case, $y = -2x + 7$).

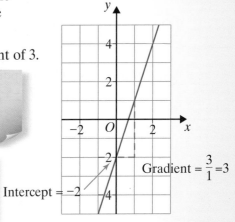

Gradient $= \dfrac{3}{1} = 3$

Intercept $= -2$

Worked Example   Problems about coordinates

Question   Where does the line $y = 7$ cross the line $y = 2x + 1$?

Solution $y = 7$ tells us that the y-coordinate is 7

$$y = 2x + 1$$
$$7 = 2x + 1 \longleftarrow \text{Substitute } y = 7$$
$$6 = 2x \longleftarrow \text{Subtract 1 from both sides of the equation}$$
$$x = 3 \longleftarrow \text{Divide both sides by 2}$$

The lines cross at (3, 7).

EXAMINER SAYS...

Candidates often solve the equation and then stop. Remember to answer the question and give the coordinates.

Worked Example   Problems about gradients

Question   **a**   Write down the equation of the line parallel to $y = 2x - 5$, which passes through the point (0, 1).

   b   Write down the equation of the line perpendicular to $y = 3x - 2$, which passes through (0, 2).

Solution   **a**   Parallel lines have the same gradient, so the gradient is 2.

   (0, 1) is on the y-axis, so the intercept is 1.

   The equation is $y = 2x + 1$.

   b   Perpendicular lines have gradients that have a product of -1.

   As $y = 3x - 2$ has a gradient of 3, the perpendicular line has a gradient of $-\frac{1}{3}$.

   The intercept on the y-axis is at 2, so the equation is $y = -\frac{1}{3}x + 2$

GET IT RIGHT!

Parallel lines have the same gradient. **Perpendicular** lines have gradients that have a product of -1.

Worked Example   The midpoint of a line segment

Question   Write down the coordinates of the point halfway between (2, -5) and (7, 1).

Solution   The midpoint is
$$\left(\frac{2 + 7}{2}, \frac{-5 + 1}{2} \right)$$
or (4.5, -2)

The midpoint of (a, b) and (c, d) is

$$\left(\frac{a + c}{2}, \frac{b + d}{2} \right)$$

The mean of the x-coordinates   The mean of the y-coordinates

EXAMINER SAYS...

You should always draw a rough sketch to make sure your answer is sensible.

Worked Example Coordinates in three dimensions

Question

a In the diagram, C is the point (4, 8, 5).

Write down the coordinates of A, B and D.

M is the point where diagonals OC and EA

intersect.

b Find the coordinates of M.

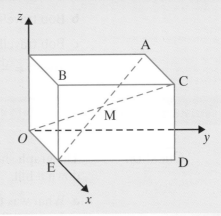

Solution

a

B, C and D have the same *x*-coordinate

A, C and D have the same *y*-coordinate.

A, B and C have the same *z*-coordinate.

C	A	B	D
(4, 8, 5)	(0, ,)	(4, ,)	(4, ,)
(4, 8, 5)	(0, 8,)	(4, 0,)	(4, 8,)
(4, 8, 5)	(0, 8, 5)	(4, 0, 5)	(4, 8, 0)

O is (0, 0, 0),
C is (4, 8, 5)

b M is the midpoint of OC, or $\left(\dfrac{0+4}{2}, \dfrac{0+8}{2}, \dfrac{0+5}{2} \right) = (2, 4, 2.5)$

Worked Example Distance-time graphs

Question

The graph shows Bob's journey from home to his grandmother's house.

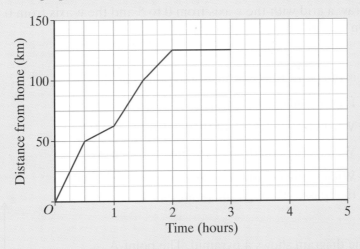

a After how long did Bob get stuck in slow-moving traffic?

b What was his average speed during the first half hour of his journey?

c Bob arrived at his grandmother's house after 2 hours. What was his average speed over the whole journey?

Solution

a After half an hour Bob got stuck in slow-moving traffic – the gradient shows he went slowly.

b Bob travelled 50 km in half an hour, which is 100 km/h.

c Bob travelled 125 km in 2 hours.

Speed $= \dfrac{\text{distance}}{\text{time}} = 62.5$ km/h

Worked Example   Speed-time graphs

Question

The graph shows the speed of a ball rolling down a hill.

a What was the speed of the ball after 5 seconds?

b What was the acceleration of the ball?

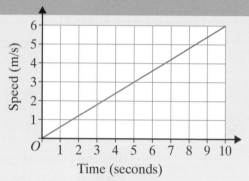

Solution

a 3 m/s

b The ball has gained 6 m/s speed in 10 seconds.

So the acceleration is $\dfrac{\text{change of speed}}{\text{time}} = \dfrac{6}{10} = 0.6$ m/s^2

GET IT RIGHT!

Be careful with units – some questions expect you to include the correct units with your answer.

Coordinates and graphs

END OF CHAPTER QUESTIONS

Time Yourself!

Can you complete these questions in **20** minutes?

1 Draw a grid with the *x*-axis from 0 to 5 and the *y*-axis from 0 to 15
On the grid, draw the graph of $y = 2x + 5$

2 A is the point (0, 4) and B is the point (2, 0).

a Find the midpoint of AB.

b What is the gradient of the line through AB?

c What is the equation of the line through AB?

d What is the equation of the line parallel to AB that passes through the point (0, −1)?

3 The diagram shows a pyramid. The point A has coordinates (3, 2, 5) and is directly over the centre of the rectangular base.
What are the coordinates of the point B?

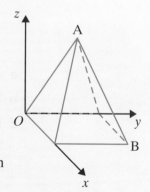

4 Find the equation of the straight line that passes through (−2, 4) and (0, 8).

3 Quadratic functions

AQA

Key words

quadratic equation

quadratic formula

discriminant

completing the square

Key points

◎ **Quadratic functions** have an x^2-term, but no higher power of x.

◎ Graphs of quadratic functions are always ∪-shaped (when the x^2-term is positive) or ∩-shaped (when the x^2-term is negative).

◎ The **quadratic formula** for $ax^2 + bx + c = 0$ is $x = \dfrac{-b \pm \sqrt{b^2 - 4ac}}{2a}$

The **discriminant** is part of that formula, $b^2 - 4ac$

◎ **Completing the square** means writing a quadratic expression in the form $(x + a)^2 + b$

Worked Example Graphs of quadratic functions

Question

a Draw the graph of $y = 2x^2 - 3x - 2$ for values of x from -3 to 4.

b Use the graph to find the solutions of $2x^2 - 3x - 2 = 0$.

c i Find the x-coordinates of the points where the curve crosses $y = 10$.

ii Write down the quadratic equation whose solutions are the answers to part **i**.

GET IT RIGHT!

Don't forget that $2x^2$ means $2 \times x \times x$. Don't work out $2x$ and then square the answer.

Solution

a

x	-3	-2	-1	0	1	2	3	4
$2x^2$	18	8	2	0	2	8	18	32
$-3x$	9	6	3	0	-3	-6	-9	-12
-2	-2	-2	-2	-2	-2	-2	-2	-2
$y = 2x^2 - 3x - 2$	25	12	3	-2	-3	0	7	18

The table breaks the equation into three parts, $2x^2$, $-3x$ and -2. These three parts are added together to calculate y.

AQA

EXAMINER SAYS...

Many candidates misuse their calculators to work out x^2 when x is negative. Some calculators give $2 \times -3^2 = -18$ instead of the correct answer of 18. You must remember to use brackets with negative numbers:
$2 \times (-3)^2 = 18$

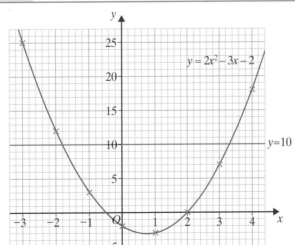

Note the scales: each small square on the x-axis is 0.2, but on the y-axis is 1

b The solutions of $2x^2 - 3x - 2 = 0$ are found where $y = 0$, which is the x-axis. From the graph, these values are $x = -0.5$ and $x = 2$

c i Drawing $y = 10$ on the graph shows the x-coordinates to be -1.8 and 3.3

ii The graphs of $y = 2x^2 - 3x - 2$ and $y = 10$ cross where $2x^2 - 3x - 2 = 10$,

or (subtracting 10 from both sides) where $2x^2 - 3x - 12 = 0$

Worked Example ## Solving quadratic equations by factorisation

(Question) Solve the equation $2x^2 - 3x = 2$

(Solution)

$$2x^2 - 3x = 2$$

$$2x^2 - 3x - 2 = 0 \longleftarrow \text{Always collect terms on one side}$$

$$(2x + 1)(x - 2) = 0 \longleftarrow \text{Factorise}$$

$$2x + 1 = 0 \quad \text{or} \quad x - 2 = 0 \longleftarrow \text{If } a \times b = 0, \text{ then either } a = 0 \text{ or } b = 0$$

$$2x = -1 \quad \text{or} \quad x = 2 \longleftarrow \text{Solve the two equations}$$

$$x = -\frac{1}{2} \quad \text{or} \quad x = 2$$

Worked Example ## Solving quadratic equations by using the formula

(Question) Solve the equation $2x^2 - 4x - 1 = 0$, giving your answers correct to 2 decimal places.

(Solution)

$$2x^2 - 4x - 1 = 0$$

$a = 2, b = -4, c = -1$ \quad Take care to include signs.

$$x = \frac{-b \pm \sqrt{b^2 - 4ac}}{2a}$$

$$x = \frac{-(-4) \pm \sqrt{(-4)^2 - 4 \times 2 \times (-1)}}{2 \times 2}$$

Using brackets on your calculator will help you avoid mistakes.

$$x = \frac{4 \pm \sqrt{(16 + 8)}}{4}$$

$$x = \frac{4 + 4.898979..}{4} \text{ or } \frac{4 - 4.898979..}{4}$$

$$x = 2.224... \text{ or } x = -0.224...$$

$$x = 2.22 \text{ or } x = -0.22 \text{ (2 d.p.)}$$

A/A*

THE GRADE To get a grade **A/A*** check your answers in the original equation (although rounding will mean the function will be very close to 0 rather than exactly 0). Candidates often make mistakes over signs and fail to give the answers to the required degree of accuracy.

Worked Example — Solving quadratic equations by completing the square

Question

a Write $x^2 - 4x - 1$ in the form $(x + a)^2 + b$

b Use your answer to part **a** to solve the equation $x^2 - 4x - 1 = 0$

Solution

a $x^2 - 4x - 1$ ← Coefficient of $x = -4$

$(x - 2)^2 = x^2 - 4x + 4$ ← Half of –4 is –2; expand $(x - 2)^2$

So $(x - 2)^2 - 5 = x^2 - 4x - 1$ ← Adjust accordingly

b $x^2 - 4x - 1 = 0$

$(x - 2)^2 - 5 = 0$ ← Substitute

$(x - 2)^2 = 5$ ← Add 5 to both sides

$x - 2 = \pm\sqrt{5}$ ← Square root both sides

$x = 2 \pm \sqrt{5}$ ← Add 2 to both sides

$x = 2 + 2.236\ldots$ or $x = 2 - 2.236\ldots$

$x = 4.24$ or $x = -0.24$ (2 d.p.)

THE GRADE To get a grade **A/A*** remember that if $y = (x + a)^2 + b$, then the minimum value is at $(-a, b)$

Quadratic functions

Time Yourself!

Can you complete these questions in **25** minutes?

END OF CHAPTER QUESTIONS

1 a Draw the graph of $y = x^2$ for values of x from –5 to 5

  b Draw the line $y = x + 1$ on the same axes.

  c Write down the x-coordinates of the points where the curve and the line cross.

  d Write down the quadratic equation whose solutions are the answers to part **c**.

2 Solve the equation $x^2 - 4x - 2 = 0$

  Give your answers in the form $p \pm \sqrt{q}$

3 a Factorise $x^2 - 5x - 24$

  b Hence solve the equation $x^2 - 5x - 24 = 0$

4 The graph of $y = x^2 - 2x + 7$ is to be used to solve the equation $x^2 - 4x + 3 = 0$

  What straight line would you need to draw?

5 Solve these equations using the stated method:

  a $x^2 - 2x - 4 = 0$ by completing the square

  b $3x^2 + 5x - 2 = 0$ by factorising

  c $2x^2 - x - 2 = 0$ by the formula

Key words

cubic function

reciprocal function

exponential function

translation

reflection

stretch

Key points

⊙ A **cubic function** has an x^3-term and has the form $y = ax^3 + bx^2 + cx + d$

⊙ A **reciprocal function** has the form $y = \dfrac{a}{x^n}$, where n is a positive integer, for example, $y = \dfrac{3}{x}, y = \dfrac{1}{x^2}$

⊙ **Exponential functions** have the form $y = Ak^{mx}$, where A, k and m are constants. For example, if £2000 is invested at 3% compound interest, the amount in the bank, £y, after x years, is given by $y = 2000 \times 1.03^x$

⊙ In the diagram:

black is **reflected** onto red
black is **translated** onto blue
black is **stretched** onto green.

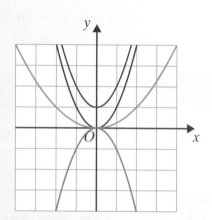

Worked Example Cubic equations and their graphs

Question   Draw the graph of $y = x^3 - 2x^2 - 1$, taking values of x from -2 to 4

Solution

GET IT RIGHT!

All cubics have the same basic shape. They are smooth and have rotation symmetry. Check any values that appear to be incorrect.

x	-2	-1	0	1	2	3	4
x^3	-8	-1	0	1	8	27	64
$-2x^2$	-8	-2	0	-2	-8	-18	-32
-1	-1	-1	-1	-1	-1	-1	-1
$y = x^3 - 2x^2 - 1$	-17	-4	-1	-2	-1	8	31

EXAMINER SAYS...

Candidates frequently forget that cubic terms can be positive or negative, whereas squared terms are always positive (or zero).

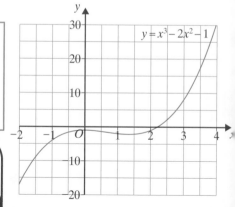

Worked Example Graphs of reciprocal functions

Question

Draw the graph of $y = \frac{2}{x}$, taking values of x from –3 to 3

Solution

x	–3	–2	–1	–0.5	–0.1	0.1	0.5	1	2	3
$x = \frac{2}{x}$	–0.67	–1	–2	–4	–20	20	4	2	1	0.67

(Decimals given to 2 d.p.)

AQA EXAMINER SAYS...

Candidates often try to join the two parts of the graph together. y is not defined when $x = 0$, so the graph is **discontinuous**.

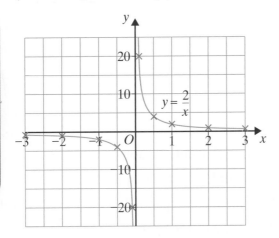

Worked Example Graphs of exponential functions

Question

a Draw the graph of $y = 2^x$, taking values of x from –2 to 4

b Use your graph to solve the equation $2^x = 7$

Solution

a

x	–2	–1	0	1	2	3	4
$y = 2^x$	0.25	0.5	1	2	4	8	16

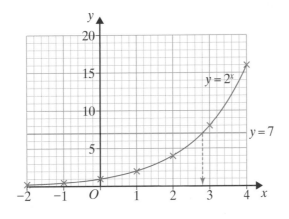

b From the graph, $y = 2^x$ and $y = 7$ cross at $x = 2.8$

Worked Example Transforming graphs

Question

The diagram shows the graph of $y = x^3$ in black and four other graphs. Match the coloured graphs to the following equations:

a $y = x^3 + 2$

b $y = (x + 2)^3$

c $y = 2x^3$

d $y = (2x)^3$

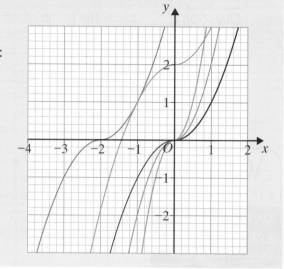

Solution

a $y = x^3 + 2$ is a translation of $\binom{0}{2}$ – blue graph

b $y = (x + 2)^3$ is a translation of $\binom{-2}{0}$ – red graph

c $y = 2x^3$ is a stretch by a scale factor 2 parallel to the y-axis – purple graph

d $y = (2x)^3$ is a stretch by a scale factor $\frac{1}{2}$ parallel to the x-axis – green graph

BUMP UP
A/A*
THE GRADE
To get a grade **A/A*** you should practise sketching graphs of a function $f(x)$ under different transformations, such as $f(x) + a$, $f(x + a)$, $af(x)$ and $f(ax)$.

Further graphs

END OF CHAPTER QUESTIONS

Time Yourself!

Can you complete these questions in **15** minutes?

1 Draw the graph of $y = 3^x$, taking values of x from -2 to 3

2 Suggest possible equations for these graphs.

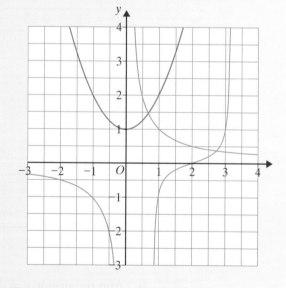

3 Draw suitable graphs in the range $-2 \leqslant x \leqslant 5$ to determine the number of solutions to the equation $x^2 + x = 2^x$

5 Simultaneous equations and inequalities

Key points

◎ An **inequality** is a statement that two expressions are not equal, for example, $n \neq 6$, $a \leqslant 4$ or $b > 3$

◎ **Simultaneous equations** are two equations that have a common solution.

Worked Example Inequalities

Question Solve $-3 < 2x + 1 \leqslant x + 5$

Solution

$-3 < 2x + 1$	and	$2x + 1 \leqslant x + 5$
$-4 < 2x$	and	$2x \leqslant x + 4$
$-2 < x$	and	$x \leqslant 4$

Write as two separate inequalities. Solve each one as if solving an equation.

$$-2 < x \leqslant 4$$

Put the inequalities back together.

THE GRADE To get a grade **A*** you should look out for questions that ask for **integer solutions**. In the above example, the integer solutions are –1, 0, 1, 2, 3, 4

Worked Example Inequalities and regions

Question Show the region defined by the inequalities $x \geqslant -2$, $y < 3$, and $y > 2x + 2$

Solution

EXAMINER SAYS...

Make sure you identify which the required region is.

The lines $y = 3$ and $y = 2x + 2$ are dotted, as the inequalities do not include the line.
x can be equal to –2, so the line $x = -2$ is solid.
The required region is the **unshaded** part.

Required region

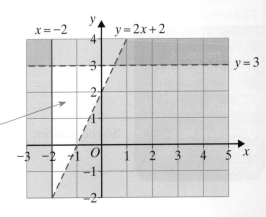

Worked Example — Solving simultaneous equations by eliminating a variable

Question

Solve these simultaneous equations:

$$3x + 2y = 9$$
$$2x - 4y = 14$$

Solution

$$3x + 2y = 9 \quad ①$$
$$2x - 4y = 14 \quad ②$$

$$① \times 2 \qquad 6x + 4y = 18 \quad ③$$
$$(+) \qquad \underline{2x - 4y = 14} \quad ②$$
$$8x \quad\ \ = 32$$
$$x = 4$$

One of the unknowns must have matching coefficients. If it hasn't, multiply one (or both) equations by finding a common multiple. In this case, the coefficients of y are made to equal 4

Substitute $x = 4$ into ①

$$3 \times 4 + 2y = 9$$
$$12 + 2y = 9$$
$$2y = -3$$
$$y = -1.5$$

As the matching terms have different signs, they are eliminated by adding. (If they had the same sign, subtracting would remove them.)

So the solution is $x = 4$ and $y = -1.5$

Worked Example — Solving simultaneous equations by substitution

Question

Solve these simultaneous equations:

$$2x - y = 9$$
$$y = x - 5$$

Solution

$$2x - y = 9 \qquad ①$$
$$y = x - 5 \qquad ②$$
$$2x - (x - 5) = 9 \longleftarrow \text{Substitute } y = x - 5 \text{ into equation } ①$$
$$2x - x + 5 = 9 \longleftarrow \text{Remove brackets, taking care with signs}$$
$$x = 4$$

Substitute $x = 4$ into equation ② .

$$y = x - 5 \qquad ②$$
$$y = 4 - 5 = -1$$

Check the answer in equation ①: $2 \times 4 - (-1) = 9$

So the solution is $x = 4$ and $y = -1$

EXAMINER SAYS...

Candidates often make mistakes when subtracting negative terms, for example, $x - 5$.

Using brackets can help to avoid errors.

A/A*

THE GRADE To get a grade **A/A*** you should familiarise yourself with the substitution method. It is particularly useful when finding intersections of graphs, as equations are often given in the form $y = ...$, ready for substitution.

Worked Example Solving simultaneous equations (non-linear)

Question

Solve these simultaneous equations:

$$y = 2x + 3$$
$$x^2 + y^2 = 5$$

Solution

$$y = 2x + 3 \qquad \text{①} \quad \text{Linear equation}$$
$$x^2 + y^2 = 5 \qquad \text{②} \quad \text{Non-linear equation}$$

$$x^2 + (2x + 3)^2 = 5 \longleftarrow \text{Substitute the linear value of } y$$
$$\text{into the non-linear equation.}$$
$$x^2 + (2x + 3)(2x + 3) = 5$$
$$x^2 + 4x^2 + 12x + 9 = 5$$
$$5x^2 + 12x + 4 = 0 \qquad \text{Solve the resulting quadratic equation.}$$
$$(5x + 2)(x + 2) = 0$$
$$5x + 2 = 0 \text{ or } x + 2 = 0$$
$$5x = -2 \text{ or } x = -2$$
$$x = -0.4 \text{ or } x = -2$$

GET IT RIGHT!

Take care when expanding terms such as $(2x + 3)^2$. Writing them in full helps to avoid errors.

Substitute into the linear equation:

When $x = -0.4$, $y = 2x + 3 = -0.8 + 3 = 2.2$

When $x = -2$, $y = 2x + 3 = -4 + 3 = -1$

So the solutions are $x = -0.4$ and $y = 2.2$ or $x = -2$ and $y = -1$

Simultaneous equations and inequalities

END OF CHAPTER QUESTIONS

Time Yourself!

Can you complete these questions in **15** minutes?

1 Solve $4x + 3 < 1$

2 Solve these simultaneous equations:
$$2x - 3y = 7$$
$$4x + y = 0$$

3 Draw a pair of axes taking values of x and y from 0 to 8

Show the region defined by the inequalities $x > 1$, $x + y \leqslant 6$ and $y > 2x$

4 Solve the simultaneous equations
$$x^2 + y = 15$$
$$x + y = 3$$

6 Proof

sum

product

consecutive

counter example

proof

Key points

◎ The **sum** of numbers is the result of adding them together.

◎ The **difference** of two numbers is the result of subtracting one number from the other.

◎ The **product** is the result of multiplying numbers.

◎ **Consecutive numbers** are next to each other, for example, 3, 4, 5.

◎ A **counter example** is an example that disproves a statement.

◎ A **proof** is a logical explanation that shows something must be true.

Worked Example Proof

GET IT RIGHT!

Negative numbers, zero and fractions (or decimals) often provide counter examples for statements that appear to be true.

(Question) a and b are both less than 10.
Is it true that $a \times b$ must be less than 100?

(Solution)   If $a = -12$ and $b = -9$, then a and b are both less than 10, but $a \times b = 108$
This is a counter example, so the statement is false.

Worked Example Algebraic proofs

(Question)

Brian writes down the number	74
He reverses the digits to get	47
He subtracts to get	27, which is 3×9

He repeats this with different two-digit starting numbers.
He notices that the answer is always a multiple of 9.
Prove that this is always true.

(Solution)

EXAMINER SAYS...

Candidates need to recognise that a two-digit number ab represents $10a + b$, and that even numbers can be represented by $2n$, and odd numbers by $2n + 1$.

74 has a value of 7 tens + 4 units.

So a two-digit number, written as ab, has a value of $10a + b$

Reversing it gives ba, which has a value of $10b + a$.

Subtracting gives $(10a + b) - (10b + a)$

$$= 10a + b - 10b - a$$

$$= 9a - 9b$$

$$= 9(a - b), \text{ which is a multiple of 9}$$

Worked Example Odd, even and multiples

(Question) Explain why the difference between the squares of two consecutive numbers is always odd.

(Solution) Let the consecutive numbers be a and $a + 1$

The difference between the squares is $(a + 1)^2 - a^2$

$$= (a + 1)(a + 1) - a^2$$
$$= a^2 + 2a + 1 - a^2$$
$$= 2a + 1$$

Since $2a$ is even, $2a + 1$ must be odd.

THE GRADE To get a grade **A/A*** you should practise algebraic proofs as these are often included on the examination paper and can be worth quite a few marks.

Proof

END OF CHAPTER QUESTIONS

Time Yourself!

Can you complete these questions in **15** minutes?

1 a Show that the sum of three consecutive numbers is always a multiple of 3.

  b Show that the sum of three consecutive even numbers is always a multiple of 6.

2 Is it true that all prime numbers can be written as the sum of two consecutive numbers?

3 Niranjan is investigating square numbers.

He notices that $7^2 = 49$

But $6 \times 8 = 48$, or (one less than 7) × (one more than 7) = (one less than 7^2)

Does this only work for 7, or will it work for any number?

1 Factorise completely, $9x^2 - 6x$ *(2 marks)*

2 Simplify $\dfrac{3x^2 + 5x - 2}{x^2 - 4}$ *(4 marks)*

3 Make p the subject of the formula $s = \dfrac{p}{4} - 2$ *(2 marks)*

4 Solve the equation $\dfrac{2x}{3} - \dfrac{x}{2} = 3$ *(3 marks)*

5 A line passes through the points A $(0, 2)$ and B $(-3, 11)$

  (a) Calculate the gradient of the line. *(2 marks)*

  (b) Write down the equation of the line. *(1 mark)*

  (c) Write down the equation of the line that passes through the point $(0, 2)$ but is perpendicular to the original line. *(2 marks)*

6 Solve the equation $x^2 + 4x - 12 = 0$ *(3 marks)*

7 Find the values of a and b such that $x^2 - 6x + 4 = (x - a)^2 + b$ *(2 marks)*

8 Solve the equation $x^2 - 4x - 2 = 0$

Give your answers to 2 decimal places. *(3 marks)*

9 Solve the equation $\dfrac{5}{x - 2} + \dfrac{x}{x + 2} = 2$ *(5 marks)*

10 (a) Draw the graph of $y = x^3 - x$, taking values of x from -3 to 3 *(3 marks)*

  (b) Use your graph to solve the equation $x^3 - x = 2$ *(1 mark)*

11 Draw the graph of $\dfrac{1}{x^2}$, taking values of x from -3 to 3 *(3 marks)*

12 Solve the simultaneous equations

    $3x + 2y = 5$

    $4x + 3y = 6$ *(4 marks)*

13 Prove that when an even number is squared, the answer is always a multiple of 4 *(3 marks)*

1 Expand and simplify $x(3x - 1) - 3(x - 5)$

The second mark is for simplifying your answer.

The first mark is for removing the brackets →

$$3x^2 - x - 3x + 15$$

$$= 3x^2 - 4x + 15$$

Answer .... *(2 marks)*

2 Factorise fully $2x^2 - 50y^2$

The first mark is awarded for recognising the common factor of 2

$$= 2(x^2 - 25y^2)$$

2 marks for factorising the difference of two squares

Answer $= 2(x + 5y)(x - 5y)$ .... *(3 marks)*

3 Make n the subject of the formula $x = \sqrt{n} + 4$

$$x - 4 = \sqrt{n}$$ The first mark is for subtracting 4.

$$(x - 4)^2 = n$$ The second mark is for squaring both sides.

Answer $n = (x - 4)^2$ .... *(2 marks)*

4 Solve the equation $x^2 - 3x + 1 = 0$, giving your answers correct to 2 decimal places.

$$a = 1, b = -3, c = 1$$

$$x = \frac{-(-3) \pm \sqrt{(-3)^2 - 4 \times 1 \times 1}}{2 \times 1}$$ The first mark is awarded for correct substitution.

$$x = \frac{3 \pm 2.236....}{2}$$ A second mark is awarded for evaluating the square root correctly.

Two more marks are awarded for the correct solutions rounded to 2 d.p.

Answer $x = 2.62$ or $x = 0.38$ (2 d.p.) .... *(4 marks)*

5 List the integer solutions to the inequality

$5 \leqslant 2n - 1 < 13$

$$5 \leqslant 2n - 1 \quad \text{and} \quad 2n - 1 < 13$$ Split into separate inequalities.

$$6 \leqslant 2n \quad \text{and} \quad 2n < 14$$ The first mark is for simplifying.

$$3 \leqslant n \quad \text{and} \quad n < 7$$ Solve each inequality for the second mark.

Answer Integer solutions are $n = 3, 4, 5,$ or 6 Give integer solutions for the third mark. *(3 marks)*

Section 1: Handling data

1 Collecting data

1 a The number of people at a rugby match is quantitative as it can be counted.

b How many tins of beans a shop sells is quantitative as it can be counted.

c The flavour of the beans is qualitative.

d The time it takes to travel from London to Manchester is quantitative as it can be measured.

2 a The number of votes for a party at a local election is discrete.

b The number of beans in a tin is discrete.

c The weight of a tin of beans is continuous as weight is a measurement and is always continuous.

d The time taken to complete this chapter is continuous as time is a measurement and is always continuous.

3 a There are $3 + 2 + 1 + 2 = 8$ red cars

b There are $2 + 4 + 3 + 0 = 9$ Vauxhall cars

c Number of black cars = 7
Total number of cars = 35
Percentage = $\frac{7}{35} \times 100 = 20\%$

4 a Assign a number to each student in the population and generate random numbers to choose a random sample of 50 students.

b Assign each student in the population a number and generate a random number to start, then sample every nth student where n = population size divided by 50

5 Completing the information for each battery in the table:

Battery	AAA	AA	9V	C	D
Number of batteries	3500	6000	700	400	400
Fraction of all batteries	$\frac{3500}{11000}$	$\frac{6000}{11000}$	$\frac{700}{11000}$	$\frac{400}{11000}$	$\frac{400}{11000}$
For a sample size of 100	$\frac{3500}{11000} \times 100 = 31.8$	$\frac{6000}{11000} \times 100 = 54.5$	$\frac{700}{11000} \times 100 = 6.4$	$\frac{400}{11000} \times 100 = 3.6$	$\frac{400}{11000} \times 100 = 3.6$
Rounded number	32	54	6	4	4

GET IT RIGHT!
Remember to round off the answers to whole numbers and check that the total does add to the sample size of 100

2 Statistical measures

1 Mean = $[19 + 10 + 17 + 18 + 19 + 18 + 17 + 22 + 7 + 13 + 18] \div 11 = 178 \div 11 = 16.2$ (3 s.f.)
For the median arrange the numbers in order:
7 10 13 17 17 18 18 18 19 19 22
Median = 18
Mode = 18
Range = $22 - 7 = 15$

2 Total age for 5 people = $5 \times 36 = 180$
Total age for 6 people = $6 \times 35 = 210$
The sixth person is $210 - 180 = 30$ years old

3 Median = 8th item = 4
Mode = 4
Mean = $64 \div 15 = 4.27$ (3 s.f.)
Range = $8 - 2 = 6$

4 a Modal class = 6–8 min

b 5.5 to 8.5

c Constructing the table for the mean:

Time (nearest min.)	Number of patients (t)	Midpoint (x)	Midpoint × frequency (fx)
3–5	43	4	172
6–8	51	7	357
9–11	27	10	270
12–14	11	13	143
15–17	5	16	80
18–22	2	20	40
23–27	0	25	0
28–40	1	34	34
Total	$\sum f = 140$		$\sum fx = 1096$

Mean = $1096 \div 140 = 7.83$ (3 s.f.)

d The 6–8 class interval contains the median.

3 Representing data

1 a

Year	2006	2006	2006	2006	2007	2007	2007	2007
Quarter	1st	2nd	3rd	4th	1st	2nd	3rd	4th
Cost	£250	£140	£90	£105	£235	£145	£45	£80
Four point moving average			£146.25	£142.50	£143.75	£132.50	£126.25	

b The four-point moving averages can be plotted on the graph as shown:

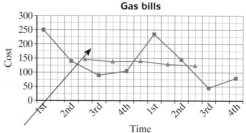

GET IT RIGHT!
The first four-point moving average is plotted in the 'middle' of the first four points.

c The trend is downwards.

2 a Completing an additional column for the cumulative frequency:

Weight (kg)	Frequency	Cumulative frequency
1–2	21	21
2–4	14	35
4–8	7	42
8–15	3	45

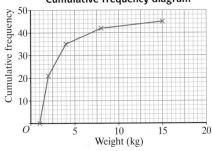

Cumulative frequency diagram

b i Median = 2.2
ii Interquartile range = 4.0 – 1.5 = 3.5

c

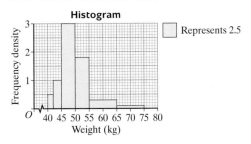

Weights of parcels

3 Completed table to show the information on the box plot:

	Years
Minimum age	0
Maximum age	70
Lower quartile age	29
Upper quartile	61
Median	42

4 a i

Weight (kg)	40⩽w<42	42⩽w<45	45⩽w<50	50⩽w<55	55⩽w<65	65⩽w<75
Frequency	1	3	15	12	3	1
Width of bar	2	3	5	5	10	10
Frequency density	0.5	1	3	2.4	0.3	0.1

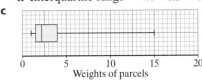

Histogram
Represents 2.5

ii

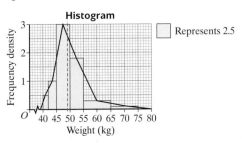

Frequency polygon
Represents 2.5

b Since the area under the graph represents the frequency then the median divides this area into two. The total area = 32 square units so draw the median line so that there are 16 square units to the left and 16 square units to the right.

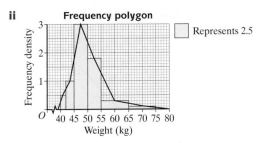

Histogram
Represents 2.5

Median = 49 kg

4 Scatter graphs

1 a Weak negative correlation
b Weak negative correlation
c No correlation
d Strong positive correlation
e No correlation
f Strong positive correlation

2 a

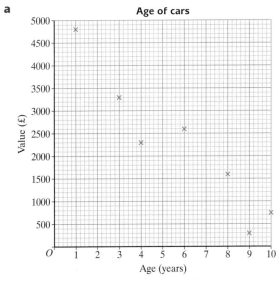

Age of cars

b The scatter graph shows moderate negative correlation.

3 a

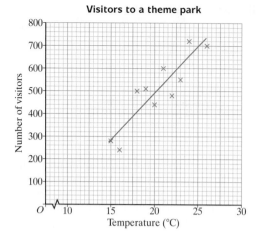

Visitors to a theme park

b i 360 people
ii 12°C
c The temperature is more likely to be inaccurate because it goes outside the given range of values.

5 Probability

1 a Relative frequency of getting a club $= \frac{11}{60}$
b Relative frequency of getting a red card $= \frac{27}{60}$ (there are 27 red cards)
c Theoretical probability of getting a spade $= \frac{1}{4}$

2 $1 - 0.85 = 0.15$

3

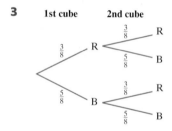

1st cube   2nd cube

a $P(R, R) = \frac{3}{8} \times \frac{3}{8} = \frac{9}{64}$
b $P(B, B) = \frac{5}{8} \times \frac{5}{8} = \frac{25}{64}$
c $P(R, B) + P(B, R) = \frac{3}{8} \times \frac{5}{8} + \frac{5}{8} \times \frac{3}{8} = \frac{30}{64} = \frac{15}{32}$

4

1st cube   2nd cube

a $P(R, R) = \frac{3}{8} \times \frac{2}{7} = \frac{6}{56}$
b $P(B, B) = \frac{5}{8} \times \frac{4}{7} = \frac{20}{56}$
c $P(R, B) + P(B, R) = \frac{3}{8} \times \frac{5}{7} + \frac{5}{8} \times \frac{3}{7} = \frac{30}{56} = \frac{15}{28}$

5

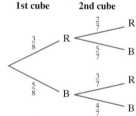

a P(not oversleep and is on time for school)
$= 0.8 \times 0.9 = 0.72$
b P(late for school)
$= 0.2 \times 0.7 + 0.8 \times 0.1$
$= 0.14 + 0.08 = 0.22$

GET IT RIGHT!

Be careful in your work with decimals. Remember 0.8×0.1 $= 0.08$ **not** 0.8

6

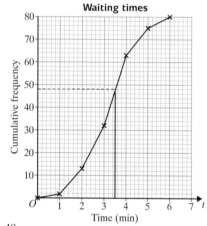

1st card   2nd card

a P(both cards are diamonds) $= \frac{1}{4} \times \frac{1}{4} = \frac{1}{16}$
b P(neither card is a diamond) $= \frac{3}{4} \times \frac{3}{4} = \frac{9}{16}$
c P(just one card is a diamond) $= \frac{1}{4} \times \frac{3}{4} + \frac{3}{4} \times \frac{1}{4} = \frac{6}{16} = \frac{3}{8}$

Section 1: Handling data

ANSWERS TO EXAMINATION STYLE QUESTIONS

1 Total number of bulbs $= 1070$
Fraction of 60 W bulbs $= \frac{680}{1070}$
Number of bulbs $= \frac{680}{1070} \times 50 = 31.775...$ rounded to 32

2 a Cumulative frequency column 32, 63, 75, 80
b

Waiting times

c 48

3 $1 \text{ cm}^2 = 2.5$ people
Number not accepted $= 2.5 \times 2.4 = 6$

4 Trend line read at 72
$(79 + 70 + 48 + x) \div 4 = 72$
$x = 91$
Costs for December $= £91$

5 a $475 \times 0.6 = 285$
b $425 \times 0.48 = 204$ boys study French
Total $= 285 + 204 = 489$

Section 2: Number

1 Integers and rounding

1 a Common factors are 1, 2, 4, 8
b Common factors are 1, 2, 3, 4, 6, 12

2 a Factors of 20 = 1, 2, 4, 5, 10, 20
Factors of 36 = 1, 2, 3, 4, 6, 9, 12, 18, 36
HCF = 4
b Factors of 30 = 1, 2, 3, 5, 6, 10, 15, 30
Factors of 45 = 1, 3, 5, 9, 15, 45
HCF = 15
c Factors of 48 =1, 2, 3, 4, 6, 8, 12, 16, 24, 48
Factors of 72 = 1, 2, 3, 4, 6, 8, 9, 12, 18, 24, 36, 72
HCF = 24

3 a Multiples of 6 = 6, 12, 18, 24, 30, …
Multiples of 15 = 15, 30, …
LCM = 30
b Multiples of 12 = 12, 24, 36, 48, 60,…
Multiples of 20 = 20, 40, 60, …
LCM = 60
c Multiples of 20 = 20, 40, 60, 80, 100, 120, 140, 160, 180, …
Multiples of 36 = 36, 72, 108, 144, 180, …
LCM = 180

4 a

2	28
2	14
7	7
	1

$28 = 2^2 \times 7$

b

2	54
3	27
3	9
3	3
	1

$54 = 2 \times 3^3$

c

2	90
3	45
3	15
5	5
	1

$90 = 2 \times 3^2 \times 5$

d

2	120
2	60
2	30
3	15
5	5
	1

$120 = 2^3 \times 3 \times 5$

e

2	250
5	125
5	25
5	5
	1

$250 = 2 \times 5^3$

f

3	567
3	189
3	63
3	21
7	7
	1

$567 = 3^4 \times 7$

5 a $\frac{1}{8}$ **b** 3 **c** $\frac{1}{0.4} = \frac{10}{4} = \frac{5}{2} = 2\frac{1}{2}$ or 2.5
d $\frac{8}{5} = 1\frac{3}{5}$ or 1.6 **e** $-\frac{1}{4}$

6 a i $72 = 2^3 \times 3^2$ **ii** $270 = 2 \times 3^3 \times 5$
b The index forms both contain 2 and 3^2 so
HCF = $2 \times 3^2 = 18$
c The LCM must include 2^3, 3^3 and 5, so the
LCM = $2^3 \times 3^3 \times 5 = 4 \times 270 = 1080$

7 a $\frac{206 \times 4.91}{1.92^2} \approx \frac{200 \times 5}{2 \times 2} = \frac{1000}{4} = 250$
b $\frac{5.74 + 3.12}{0.879} \approx \frac{6 + 3}{0.9} = \frac{9}{0.9} = \frac{90}{9} = 10$
c $\frac{83.9}{19.3 \times 0.163} \approx \frac{80}{20 \times 0.2} = \frac{80}{4} = 20$
d $\frac{86.71 - 32.04}{41.56 + 79.52} \approx \frac{90 - 30}{40 + 80} = \frac{60}{120} = 0.5$

8 a Maximum possible total weight = 15.5 + 20.5 = 36 kg
b Maximum possible difference = 20.5 – 14.5 = 6 kg

9 Upper bound = $\frac{4.65 \times 3.25}{2} = 7.55625$ cm^2

Lower bound = $\frac{4.55 \times 3.15}{2} = 7.16625$ cm^2

10 Maximum time = $\frac{805}{29.5} = 27.3$ minutes

Minimum time = $\frac{795}{30.5} = 26.1$ minutes

2 Decimals, fractions and surds

1 a $48.52 + 6 \div 10 = 48.52 + 0.6 = 49.12$
b $3.2 \times 100 - 0.84 \times 10 = 320 - 8.4 = 311.6$
c $0.4 \times (5 - 1.3) = 0.4 \times 3.7 = 1.48$
d $\frac{0.2 \times 1.4}{1.2 - 0.85} = \frac{0.28}{0.35} = \frac{28}{35} = \frac{4}{5} = 0.8$

Make sure you can also do these on a calculator.

2 $\frac{3}{5} = \frac{24}{40}$ $\frac{7}{10} = \frac{28}{40}$ $\frac{1}{2} = \frac{20}{40}$ $\frac{5}{8} = \frac{25}{40}$ (nearest)
$\frac{3}{4} = \frac{30}{40}$ $\frac{13}{20} = \frac{26}{40}$

3 $\frac{2}{3} = 0.\dot{6}, \frac{5}{9} = 0.\dot{5}, \frac{3}{11} = 0.\dot{2}\dot{7}, \frac{5}{6} = 0.8\dot{3}$ and $\frac{6}{7} = 0.\dot{8}5714\dot{2}$

The others all terminate because the prime factors of their denominators are 2 and 5

4 a $x = 0.\dot{4}, 10x = 4.\dot{4}, 9x = 4, x = \frac{4}{9}$
b $x = 0.\dot{3}\dot{7}, 100x = 37.\dot{3}\dot{7}, 99x = 37, x = \frac{37}{99}$
c $x = 0.\dot{1}\dot{2}, 100x = 12.\dot{1}\dot{2}, 99x = 12, x = \frac{12}{99} = \frac{4}{33}$
d $x = 0.\dot{5}4\dot{9}, 1000x = 549.\dot{5}4\dot{9}, 999x = 549, x = \frac{549}{999} = \frac{61}{111}$

5 $2\frac{1}{2} - \frac{2}{3} = 2 + \frac{3}{6} - \frac{4}{6} = 1 + \frac{9}{6} - \frac{4}{6} = 1\frac{5}{6}$ pints

6 a $\frac{1}{3} + \frac{1}{6} = \frac{2}{6} + \frac{1}{6} = \frac{3}{6} = \frac{1}{2}$ **b** $\frac{7}{8} - \frac{3}{5} = \frac{35}{40} - \frac{24}{40} = \frac{11}{40}$
c $\frac{2}{7} \times \frac{5}{6} = \frac{5}{21}$ **d** $\frac{3}{4} \div \frac{9}{20} = \frac{3}{4} \times \frac{20}{9} = \frac{5}{3} = 1\frac{2}{3}$
e $3\frac{2}{3} + 1\frac{4}{5} = 4 + \frac{10}{15} + \frac{12}{15} = 4 + \frac{22}{15} = 4 + 1\frac{7}{15} = 5\frac{7}{15}$
f $4\frac{1}{2} - 2\frac{5}{8} = 2 + \frac{4}{8} - \frac{5}{8} = 1 + \frac{12}{8} - \frac{5}{8} = 1\frac{7}{8}$
g $2\frac{1}{2} \times 2\frac{4}{5} = \frac{5}{2} \times \frac{14}{5} = \frac{7}{1} = 7$
h $1\frac{3}{4} \div 2\frac{5}{8} = \frac{7}{4} \div \frac{21}{8} = \frac{7}{4} \times \frac{8}{21} = \frac{2}{3}$

7 $\frac{1}{3} + \frac{1}{4} = \frac{4}{12} + \frac{3}{12} = \frac{7}{12}$
$8 \times \frac{7}{12} = \frac{8}{1} \times \frac{7}{12} = \frac{14}{3} = 4\frac{2}{3}$ 5 tins are needed

8 a $\sqrt{5} \times \sqrt{10} = \sqrt{50} = \sqrt{25 \times 2} = 5\sqrt{2}$
b $\sqrt{27} + \sqrt{12} - 4\sqrt{3} = 3\sqrt{3} + 2\sqrt{3} - 4\sqrt{3} = \sqrt{3}$
c $\frac{\sqrt{8}}{4} = \frac{2\sqrt{2}}{4} = \frac{\sqrt{2}}{2}$
d $\frac{\sqrt{60}}{\sqrt{5}} = \sqrt{\frac{60}{5}} = \sqrt{12} = 2\sqrt{3}$

Other methods are possible.

9 a $(\sqrt{2} + 3\sqrt{2})^2 = (4\sqrt{2})^2 = 16 \times 2 = 32$
b $(\sqrt{5} - 1)(\sqrt{5} - 1) = 5 - \sqrt{5} - \sqrt{5} + 1 = 6 - 2\sqrt{5}$
c $(4 + \sqrt{3})(4 - \sqrt{3}) = 16 - 4\sqrt{3} + 4\sqrt{3} - 3 = 13$
d $(2 - \sqrt{5})(3 - \sqrt{5}) = 6 - 2\sqrt{5} - 3\sqrt{5} + 5 = 11 - 5\sqrt{5}$

10 a $\frac{4}{\sqrt{2}} = \frac{4\sqrt{2}}{2} = 2\sqrt{2}$
b $\frac{\sqrt{7}}{\sqrt{3}} = \frac{\sqrt{7} \times \sqrt{3}}{3} = \frac{\sqrt{21}}{3}$
c $\frac{3\sqrt{8}}{\sqrt{2}} = \frac{3\sqrt{8} \times \sqrt{2}}{2} = \frac{3\sqrt{16}}{2} = \frac{3 \times 4}{2} = 6$

d $\dfrac{1}{3 + \sqrt{2}} = \dfrac{1}{(3 + \sqrt{2})} \times \dfrac{(3 - \sqrt{2})}{(3 - \sqrt{2})}$

$= \dfrac{(3 - \sqrt{2})}{9 - 3\sqrt{2} + 3\sqrt{2} - 2} = \dfrac{3 - \sqrt{2}}{7}$

e $\dfrac{2}{3 - \sqrt{5}} = \dfrac{2}{(3 - \sqrt{5})} \times \dfrac{(3 + \sqrt{5})}{(3 + \sqrt{5})} = \dfrac{2(3 + \sqrt{5})}{9 + 3\sqrt{5} - 3\sqrt{5} - 5}$

$= \dfrac{2(3 + \sqrt{5})}{4} = \dfrac{3 + \sqrt{5}}{2}$

f $\dfrac{2 - \sqrt{3}}{2 + \sqrt{3}} = \dfrac{(2 - \sqrt{3})}{(2 + \sqrt{3})} \times \dfrac{(2 - \sqrt{3})}{(2 - \sqrt{3})} = \dfrac{4 - 2\sqrt{3} - 2\sqrt{3} + 3}{4 - 2\sqrt{3} + 2\sqrt{3} - 3}$

$= \dfrac{7 - 4\sqrt{3}}{1} = 7 - 4\sqrt{3}$

In many questions there are alternative methods.

3 Percentages and proportion

1 10% of 2.5 litres = 0.25 litres.
Small tin contains 80% = 8×0.25 = 2 litres

2 10% of £26.80 = £2.68
5% of £26.80 = £1.34
2.5% of £26.80 = £0.67
Price including VAT = £26.80 + £4.69 = £31.49

On a calculator:
1.175 × £26.80 = £31.49

3 1.043 × 366 = £381.738 = £381.74 (nearest penny)

4 Each year the value falls by a quarter.
After 1st year value = £30 000
After 2nd year value = £30 000 - £7 500 = £22 500

On a calculator:
0.75² × £40 000 = £22 500

5 1.05⁵ × £6000 = £7657.689375 = £7657.69 (nearest penny)

6 They have enough for 10×25 = 250 drinks. Percentage
of milk used = $\dfrac{235}{{}_5\cancel{250}} \times \cancel{100}^{\,2}$ = 94%

Alternatively:
319 ÷ 275 = 1.16

7 Increase = £319 000 – £275 000 = £44 000
Percentage increase = 44 ÷ 275 = 0.16 = 16%

8 65% of original price = £31.20
1% of original price = £31.20 ÷ 65 = £0.48
Original price = £48

9 R : S : T = 12 : 18 : 20 = 6 : 9 : 10
1 part = £140 ÷ 25 = £5.60
Ruth gets £5.60 × 6 = £33.60,
Sue gets £5.60 × 9 = £50.40,
Tom gets £5.60 × 10 = £56

Remember to check the total.

10 As t increases from 1 to 2 (twice as large), v increases from
0.5 to 4 (8 times as large)
This suggests $v \propto t^3$, that is, $v = kt^3$. Substituting the first pair
of values $t = 1$ and $v = 0.5$ gives $k = 0.5$
Testing the other points in $v = 0.5t^3$:
When $t = 2$, $v = 0.5 \times 2^3 = 0.5 \times 8 = 4$
When $t = 5$, $v = 0.5 \times 5^3 = 0.5 \times 125 = 62.5$
This agrees with the table, so $v \propto t^3$.

11 a $Y = kZ^2$ Substituting $Y = 72$, $Z = 3$
gives $72 = k \times 3^2$ then $k = \frac{72}{9} = 8$, so $Y = 8Z^2$
b When $Z = 5$, $Y = 8 \times 5^2 = 8 \times 25 = 200$
c When $Y = 18$, $18 = 8Z^2$ giving $Z^2 = \frac{18}{8} = \frac{9}{4}$
so $Z = \frac{3}{2} = 1\frac{1}{2}$ or 1.5   *The question says Z is positive, so it cannot be –1.5*

12 $P = \dfrac{k}{\sqrt{Q}}$ Substituting $P = 6$, $Q = 4$ gives $6 = \dfrac{k}{\sqrt{4}}$
and so $k = 12$ and $P = \dfrac{12}{\sqrt{Q}}$
When $Q = 16$, $P = \dfrac{12}{\sqrt{16}} = 3$

4 Indices and standard form

1 a $7^3 = 7 \times 7 \times 7 = 343$ **b** $6^0 = 1$
c $2^{-4} = \dfrac{1}{2^4} = \dfrac{1}{16}$ **d** $-2^4 = -(2 \times 2 \times 2 \times 2) = -16$
e $(-2)^4 = (-2) \times (-2) \times (-2) \times (-2) = 16$
f $16^{\frac{1}{2}} = \sqrt{16} = 4$ **g** $27^{\frac{2}{3}} = \sqrt[3]{27^2} = 3^2 = 9$
h $36^{-\frac{1}{2}} = \dfrac{1}{\sqrt{36}} = \dfrac{1}{6}$ **i** $4^{-\frac{3}{2}} = \dfrac{1}{\sqrt{4^3}} = \dfrac{1}{2^3} = \dfrac{1}{8}$
j $125^{-\frac{2}{3}} = \dfrac{1}{\sqrt[3]{125^2}} = \dfrac{1}{5^2} = \dfrac{1}{25}$

2 a $2^4 \times 3^2 = 2 \times 2 \times 2 \times 2 \times 3 \times 3 = 144$
b $10^3 \div 10^5 = 10^{3-5} = 10^{-2} = \dfrac{1}{10^2} = \dfrac{1}{100}$
c $(2^2)^3 = 2^2 \times 2^2 \times 2^2 = 2^{2 \times 3} = 2^6 = 64$
d $\dfrac{2^3}{3^2} = \dfrac{2 \times 2 \times 2}{3 \times 3} = \dfrac{8}{9}$

Check your answers on a calculator.

3 a $x^5 \times x^4 = x^{5+4} = x^9$
b $x^8 \div x^2 = x^{8-2} = x^6$
c $(x^6)^3 = x^6 \times x^6 \times x^6 = x^{6 \times 3} = x^{18}$
d $\dfrac{x^3 \times x}{x^7} = \dfrac{x^{3+1}}{x^7} = \dfrac{x^4}{x^7} = x^{4-7} = x^{-3} = \dfrac{1}{x^3}$

4 a 9×10^6 **b** 4×10^{-5} **c** 4.62×10^{11} **d** 5.7×10^{-8}

5 a 300 000 000   **b** 74 200 000 000
c 0.0003   **d** 0.00000000256

6 a $(5 \times 10^4) \times (8 \times 10^7) = 40 \times 10^{11} = 4 \times 10^{12}$
b $(3 \times 10^{12}) \div (4 \times 10^5) = 0.75 \times 10^7 = 7.5 \times 10^6$
c $(9 \times 10^{-4})^2 = 81 \times 10^{-8} = 8.1 \times 10^{-7}$
d $(6 \times 10^4) \div 2 = 3 \times 10^4$
e 10% of $3 \times 10^6 = 3 \times 10^5$ (÷ by 10)
5% of $3 \times 10^6 = 1.5 \times 10^5$ (half of 10%)
15% of $3 \times 10^6 = 4.5 \times 10^5$ (adding)

7 a $(3.28 \times 10^7) \times (9.67 \times 10^{-3}) = 317\,176 = 3.17 \times 10^5$ (3 s.f.)
b $(8.39 \times 10^4) \div (1.76 \times 10^{-8}) = 4.767\ldots \times 10^{12}$
$= 4.77 \times 10^{12}$ (3 s.f.)
c $(6.25 \times 10^5)^{-3} = 4.096 \times 10^{-18} = 4.10 \times 10^{-18}$ (3 s.f.)
d $(4.62 \times 10^{-5}) + (1.97 \times 10^{-6}) = 4.817 \times 10^{-5}$
$= 4.82 \times 10^{-5}$ (3 s.f.)

8 $\dfrac{4\,000\,000\,000\,000}{60\,000\,000} = \dfrac{400\,000}{6} = £70\,000$ (1 s.f.)

or £(7×10^4) in standard form.

Section 2: Number

ANSWERS TO EXAMINATION STYLE QUESTIONS

1 a

2	560
2	280
2	140
2	70
5	35
7	7
	1

$560 = 2^4 \times 5 \times 7$

b i $72 = 36 \times 2 = 2^2 \times 3^2 \times 2$
$= 2^3 \times 3^2$
ii $360 = 72 \times 5 = 2^3 \times 3^2 \times 5$
(or use $360 = 36 \times 2 \times 5$)

c HCF of 360 and 560
$= 2^3 \times 5 = 8 \times 5 = 40$

2 $\dfrac{318 \times 5.09}{0.395} \approx \dfrac{300 \times 5}{0.4} = \dfrac{1500}{0.4} = \dfrac{15\,000}{4} = 3750$

3 Minimum distance = 59.5 m   Maximum time = 12.5 s
Minimum average speed = $59.5 \div 12.5 = 4.76$ m s^{-1}

4 a $x = 0.4\dot{6}$ $100x = 46.4\dot{6}$ $99x = 46$ $x = \frac{46}{99}$

b $0.14\dot{6} = 0.1 + 0.04\dot{6} = \frac{1}{10} + \frac{46}{990}$
$= \frac{99}{990} + \frac{46}{990} = \frac{145}{990} = \frac{29}{198}$

Alternative methods are acceptable.

5 $\dfrac{2 + \sqrt{6}}{\sqrt{6}} = \dfrac{2 + \sqrt{6}}{\sqrt{6}} \times \dfrac{\sqrt{6}}{\sqrt{6}} = \dfrac{2\sqrt{6} + 6}{6}$
Cancelling by 2 gives $\dfrac{\sqrt{6} + 3}{3}$ or $\dfrac{\sqrt{6}}{3} + 1$

6 Area of rectangle A = $(\sqrt{15} - \sqrt{3})(\sqrt{15} + \sqrt{3})$
$= 15 - 3 = 12$ cm^2
Length of rectangle B = $\dfrac{12}{\sqrt{2}} = \dfrac{12\sqrt{2}}{2} = 6\sqrt{2}$ cm

7 a Price at Cut Prices = $1.175 \times £80 = £94$
This is £2 more than The Audio Store.
b Percentage decrease = $\frac{9}{75} \times 100 = 12\%$
c 40% of original price = £18.60
Original price = $\dfrac{£18.60}{0.4} = £46.50$

8 Repeatedly multiplying by 1.035 gives the following amounts
After 1 year: £4347
After 2 years: £4499.145
After 3 years: £4656.615075
After 4 years: £4819.596603
After 5 years: £4988.282484
After 6 years: £5162.872371
Amount exceeds £5000 after 6 years

9 Milk : dark = $18 : 12 = 3 : 2$ $40 \div 5 = 8$
Number of dark chocolates = $2 \times 8 = 16$

10 When t is multiplied by 3 (from 2 to 6), w is multiplied by 9
(6 to 54).
This suggests $w = kt^2$ Substituting $t = 2$, $w = 6$ gives $6 = 4k$
and $k = 1.5$ so $w = 1.5t^2$
Checking the other values in the table: When $t = 3$,
$w = 1.5t^2 = 1.5 \times 9 = 13.5$
When $t = 6$, $w = 1.5t^2 = 1.5 \times 36 = 54$
Both are correct so $w \propto t^2$ fits the results.

11 $W = \dfrac{k}{D}$ Substituting $W = 25$, $D = 16$ gives $25 = \dfrac{k}{16}$
so $k = 400$.
When $D = W$, $W = \dfrac{400}{W}$ giving $W^2 = 400$ and so $W = 20$
(since W is positive).

12 a i $6p^0 = 6 \times 1 = 6$ **ii** $(6p)^0 = 1$
b $\dfrac{1}{16} = \dfrac{1}{2^4} = 2^{-4}$, so $x = -4$
c $81^{\frac{1}{2}} = \sqrt{81} = 9 = 3^2$ and $27^y = (3^3)^y = 3^{3y}$, so $3y = 2$
and $y = \frac{2}{3}$

13 a Reciprocal of $0.8 = \dfrac{1}{0.8} = 1.25$
b i $\sqrt{6.4^2 + 3.18^3} = 8.550873172$ **ii** 8.6 (2 s.f.)

14 $A = 4\pi r^2 = 4 \times \pi \times 3480^2 = 1.52 \times 10^8$ km^2 (3 s.f.)

Section 3: Space, shape and measures

1 Area and volume

1 a Area = $10.4 \times 2.2 + (8.5 - 2.2) \times 2.2 = 36.74$ cm^2
or = $10.4 \times 8.5 - (8.5 - 2.2) \times 4.1 \times 2 = 36.74$ cm^2

b Perimeter = large semicircle + small semicircle + line
$= \dfrac{\pi \times 12}{2} + \dfrac{\pi \times 6}{2} + 6 = 9\pi + 6$ cm
$= 34.3$ cm (1 d.p.)

2 a Volume = area of cross-section × length
$= \frac{1}{2} \times$ base × height × length
$= \frac{1}{2} \times 6 \times 8 \times 3 = 72$ cm^3
b Surface area = 2 × area of triangle + area of base +
area of back + area of slope
$= 2 \times \frac{1}{2} \times 6 \times 8 + 6 \times 3 + 8 \times 3 + 10 \times 3$
$= 120$ cm^2

3 a Area of trapezium = $\frac{1}{2}(a + b)h$
$= \frac{1}{2}(1.8 + 2.5) \times 2 = 4.3$ m^2
b Volume = area of cross-section × length
$= 4.3 \times 3 = 12.9$ m^3

4 Height = volume ÷ area of base = $\dfrac{14}{7} = 2$ cm
Base must be 7 cm × 1 cm if it is made up of centimetre cubes.
Volume = 7 cm × 1 cm × 2 cm
Surface area = $(7 \times 1) \times 2 + (7 \times 2) \times 2 + (1 \times 2) \times 2$
$= 46$ cm^2

5 Surface area = area of top + area of curved surface
$= \pi r^2 + \pi r l = \pi \times 6^2 + \pi \times 6 \times 20$
$= 156\pi$ cm^2

6 Surface area = $4\pi r^2 = 222$
$r^2 = \dfrac{222}{4\pi}$
$r = \sqrt{\dfrac{222}{4\pi}} = 4.203\ldots = 4.2$ mm (1 d.p.)

7 Volume of sphere = $\frac{4}{3}\pi r^3 = \frac{4}{3}\pi \times 3^3 = 36\pi$ cm^3
Volume of cone = $\frac{1}{3}\pi r^2 h = \frac{1}{3}\pi r^2 \times 3 = \pi r^2$ cm^3
These are equal so $36\pi = \pi r^2$
$r^2 = 36$
$r = 6$ cm

8 Area of sector = $\dfrac{\text{angle AOB}}{360} \times \pi r^2 = \dfrac{\text{angle AOB}}{360} \times \pi \times 64$
$37 = \dfrac{\text{angle AOB}}{360} \times \pi \times 64$
angle AOB = $\dfrac{360 \times 37}{64 \times \pi} = 66.24\ldots = 66.2°$ (1 d.p.)

ANSWERS

2 Polygons and circles

1 angle CDA + 72° + 150° + 88° = 360° (angles in a
quadrilateral)
angle CDA + 310° = 360°
angle CDA = 360° − 310° = 50°

x + angle CDA = 180° (angles on a straight
line)
x + 50° = 180°
x = 180° − 50° = 130°

2 a angle OQR = angle OSR = 90° (angle between chord
and tangent)
angle QOS + angle OQR + angle OSR + angle QRS
= 360° (angle sum of a quadrilateral)
x + 90° + 90° + 50° = 360°
x = 360° − 230° = 130°
b angle QOS = 2 × angle QPS (angle at centre is twice
angle at circumference)
$$y = \frac{130°}{2} = 65°$$

3 angle OCD = 90° (angle between chord and tangent)
angle OCB = angle OCD − angle BCD
= 90° − 75° = 15°
angle AOC = 2 × angle ABC (angle at centre is twice angle
at circumference)
$$\text{angle ABC} = \frac{120°}{2} = 60°$$
angle AOC (reflex) = 360° − 120° = 240° (angles at a point)
angle OAB + angle ABC + angle OCB + angle AOC
(reflex) = 360° (angle sum of a quadrilateral)
x + 60° + 15° + 240° = 360°
x = 360° − 315° = 45°

4 a Let POQ be the diameter.
S is a point on the circumference.
ST is a line through O.
angle SPO = angle PSO = a
(base angles of isosceles triangle)
angle POS = 180° − 2a
(angle sum of triangle OSP)
angle QSO = angle SQO = b
(isosceles triangle)
angle PSO + angle QSO + angle SQO + angle SPO
= 180° (angle sum of triangle PSQ)
$a + b + b + a$ = 180°
so 2a + 2b = 180°
2(a + b) = 180°
(a + b) = 90°
so angle PSQ = 90°

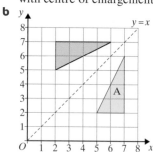

b TS is a tangent at S to the circle
centre O.
Q and R are two points on the
circumference.

angle OQS = angle OSQ = a
(base angles of isosceles
triangle)
angle QOS = 180° − 2a
(angle sum of triangle QOS)
$$\text{angle QRS} = \frac{180° - 2a}{2} = 90° - a$$
(angle at centre is twice angle at circumference)
angle OST = 90° (angle between radius and tangent)
angle QST = 90° − a
so angle QST = angle QRS

5 a angle DOC = 2 × angle DAC (angle at centre is twice
angle at circumference)
x = 2a
angle DOC = 2 × angle DBC (angle at centre is twice
angle at circumference)
x = 2b
So 2a = 2b
angle a = angle b
b 11x + 10x + 9x + 10x = 360° (angle sum of a
quadrilateral)
40x = 360°
$$x = \frac{360°}{40} = 9°$$
angle DAB = 11x = 11 × 9° = 99°
angle BCD = 9x = 9 × 9° = 81°
angle ABC = angle ADC = 10x = 10 × 9° = 90°

angle DAB + angle BCD = 99° + 81° = 180°
angle ABC + angle ADC = 90° + 90° = 180°
ABCD is a cyclic quadrilateral as both pairs of opposite
angles add up to 180°

3 Transformations

1 a Enlargement by scale factor $\frac{1}{2}$
with centre of enlargement (1, 4).
b

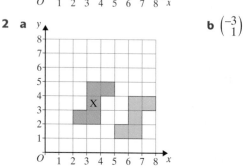

2 a

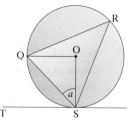

b $\begin{pmatrix} -3 \\ 1 \end{pmatrix}$

3 a Yes   **b** Yes   **c** No   **d** No

4 a   **b**

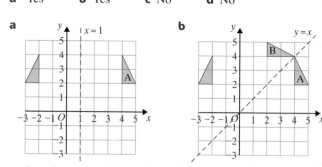

c Rotation 90° clockwise about (1, 1)

5 a Area **b** Length **c** Length **d** Volume **e** None

6

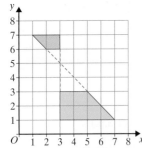

7 a i $\overrightarrow{AE} = \overrightarrow{AB} + \overrightarrow{BE}$
$= \overrightarrow{AB} - \overrightarrow{EB}$
$= 2\mathbf{a} - 2\mathbf{b}$
ii $\overrightarrow{AC} = 2\overrightarrow{AE}$
$= 2(2\mathbf{a} - 2\mathbf{b})$
$= 4\mathbf{a} - 4\mathbf{b}$

b i $\overrightarrow{DF} = \overrightarrow{DB} + \overrightarrow{BF}$
$= \frac{1}{2}\overrightarrow{AB} + \frac{1}{2}\overrightarrow{BE}$
$= \frac{1}{2}\overrightarrow{AB} - \frac{1}{2}\overrightarrow{EB}$
$= \frac{1}{2}(2\mathbf{a}) - \frac{1}{2}(2\mathbf{b})$
$= \mathbf{a} - \mathbf{b}$
ii $\overrightarrow{AE} = 2\mathbf{a} - 2\mathbf{b} = 2(\mathbf{a} - \mathbf{b})$
$= 2\overrightarrow{DF}$
Since $\overrightarrow{AE}$ is a multiple of $\overrightarrow{DF}$, AE and DF are parallel.

4 Measures and drawing

1 a   **b**

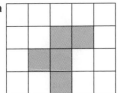   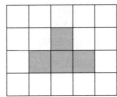

2 a minimum = 29.5 g
maximum = 30.5 g
b minimum = 6 × 29.5 = 177 g
maximum = 6 × 30.5 = 183 g

3 a Draw a straight line.
With point of compasses on P draw a large arc to cut line at Q.
With same radius and point of compasses on Q draw arc to cut other arc at R.
Join PR.
Angle RPQ = 60°.
b A is due west of B so draw a horizontal line AB = 6 cm.
With point of compasses on A, radius 4 cm, draw an arc.
With point of compasses on B, radius 5 cm, draw an arc.
Shade area where arcs cross.

4 A straight line 2 cm above and below original line joined at each end by a semicircle, radius 2 cm.

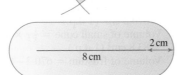

5 time $= \frac{\text{distance}}{\text{speed}}$
Ben's time $= \frac{6}{8} = \frac{3}{4}$ h = 45 min
Zoë's time $= \frac{4}{3} = 1\frac{1}{3}$ h = 80 min
Difference = 80 − 45 = 35 min

6 8.95 s ≈ 9 s
speed $= \frac{\text{distance}}{\text{time}}$
speed $= \frac{100}{9}$ m/s
$= \frac{100}{9} \times 60 \times 60$ metres/ hour
$= \frac{100 \times 60 \times 60}{9 \times 1000}$ km/h
= 40 km/h

7 Draw a long line and mark off AB = 6 cm.
Draw two arcs, centre A, equal radius, either side of A.
Centre on each arc, draw arcs that cross above A.
Join this point to A.
Measure 4 cm up this line and mark off C.
Join BC and measure.
BC = 7.2 cm

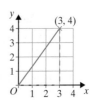

8 Lower bound of volume = 99.5 cm³
Lower bound of side $= \sqrt[3]{99.5} = 4.6338... = 4.63$ cm (3 s.f.)

9 The line from the origin to (3, 4) is a radius.
Using Pythagoras' theorem:
$x^2 + y^2 = r^2$
$3^2 + 4^2 = r^2$
$r^2 = 25$
The equation of the circle is $x^2 + y^2 = 25$

5 Pythagoras' theorem

1 $EF^2 = DF^2 + DE^2$
$13^2 = DF^2 + 12^2$
$169 = DF^2 + 144$
$25 = DF^2$
$DF = \sqrt{25} = 5$ cm   (or recognise a 5, 12, 13 Pythagorean triple)

2 a The corresponding sides are in the same ratio,
AB:DE = 2.4:3.2 = 24:32 = 3:4
$\frac{BC}{EF} = \frac{3}{4}$
$\frac{BC}{9.6} = \frac{3}{4}$
$9.6 \times \frac{BC}{9.6} = \frac{3}{4} \times 9.6$
$BC = \frac{3}{4} \times 9.6 = 7.2$ cm
b Using Pythagoras' theorem in triangle DEF,
$EF^2 = DF^2 + DE^2$
$9.6^2 = DF^2 + 3.2^2$
$92.16 = DF^2 + 10.24$
$81.92 = DF^2$
$DF = \sqrt{81.92} = 9.05... = 9.1$ cm (1 d.p.)
c angle BCA = 19°

3 In triangle ABD
$AD^2 = AB^2 + BD^2$
$15^2 = 12^2 + BD^2$
$225 = 144 + BD^2$
$81 = BD^2$
$BD = \sqrt{81} = 9$ cm (or recognise a 9, 12, 15
Pythagorean triple)
In triangle BDC
$BC^2 = BD^2 + DC^2$
$BC^2 = 9^2 + 7^2$
$BC^2 = 81 + 49$
$BC^2 = 130$
$BC = \sqrt{130}$ cm

4 a angle CAB = angle CED (given)
angle ACB = angle ECD (opposite angles)
BC = DC (given)
triangle ABC ≡ triangle EDC (AAS)
b EC = AC = 2.6 cm

5 In triangle SPQ,   In triangle WSQ,
$SQ^2 = SP^2 + PQ^2$ $WQ^2 = WS^2 + SQ^2$
$SQ^2 = 2^2 + 2^2$ $WQ^2 = 2^2 + 8$
$SQ^2 = 4 + 4 = 8$ $WQ^2 = 4 + 8 = 12$
$SQ = \sqrt{8}$ cm $WQ = \sqrt{12}$ cm $= 2\sqrt{3}$ cm

6 Linear ratio = 25 : 36

Volume ratio = $25^3 : 36^3 = 15\ 625 : 46\ 656 = 1 : 2.98...$
= 1 : 3 (1 s.f.)

÷15 625
÷15 625

The larger pack is approximately 3 times the volume of the smaller pack so the claim is justified.

6 Trigonometry

1 a $\cos 25° = \dfrac{adj}{hyp}$

$\cos 25° = \dfrac{AC}{9.4}$

$9.4 \times \cos 25° = AC$
$AC = 8.519... = 8.5$ cm (1 d.p.)

b $\tan F = \dfrac{opp}{adj}$

$\tan F = \dfrac{7}{4} = 1.75$

$F = \tan^{-1} 1.75$

angle F = 60.255... = 60.3° (1 d.p.)

2 a $\tan x = \dfrac{opp}{adj}$

$\tan x = \dfrac{5.2}{1.6} = 3.25$

$x = \tan^{-1} 3.25$
= 72.89... = 72.9° (1 d.p.)
The ladder is safe.
b The ladder reaches highest up the wall when the angle is greatest.

$\sin 78° = \dfrac{opp}{hyp}$

$\sin 78° = \dfrac{opp}{6}$

$6 \times \sin 78° = opp$

opp = 5.868... = 5.87 m (3 s.f.)
The maximum height is 5.87 m (3 s.f.)

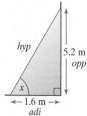
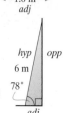

3 a Using the sine rule in triangle ABC:

$\dfrac{\sin A}{a} = \dfrac{\sin B}{b}$

$\dfrac{\sin 50°}{15} = \dfrac{\sin B}{10}$

$10 \times \dfrac{\sin 50°}{15} = \sin B$

sin B = 0.5106962954...
B = sin⁻¹ 0.5106962954...
angle ABC = 30.71... = 30.7° (1 d.p.)

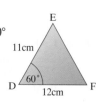

b Using the cosine rule in triangle DEF:
$EF^2 = 11^2 + 12^2 - 2 \times 11 \times 12 \times \cos 60°$
$EF^2 = 121 + 144 - 264\cos 60°$
$= 265 - 132$
$= 133$
EF = 11.53... = 12 cm (nearest cm)

4 Using the area rule:
Area $= \frac{1}{2} \times 10 \times 8 \times \sin 45°$
$= \frac{1}{2} \times 10 \times 8 \times \dfrac{\sqrt{2}}{2}$
$= 40 \times \dfrac{\sqrt{2}}{2}$
$= 20\sqrt{2}$ cm²

5 Required angle is angle EAC.
Using Pythagoras' theorem in triangle ABC,
$AC^2 = AB^2 + BC^2$
$AC^2 = 80^2 + 95^2$
$= 6400 + 9025 = 15\ 425$
$AC = \sqrt{15\ 425}$

In triangle EAC,

$\tan A = \dfrac{opp}{adj}$

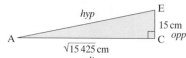

$\tan A = \dfrac{15}{\sqrt{15\ 425}} = 0.1207754523...$

$A = \tan^{-1} 0.1207754523...$
$= 6.886... = 6.9°$ (1 d.p.)
angle EAC = 6.9° (1 d.p.)

6 a $x = 360° - 75° = 285°$
b $x = 180° - 75° = 105°$ and
$x = 180° + 75° = 255°$

Section 3: Shape, space and measures
ANSWERS TO EXAMINATION STYLE QUESTIONS

1 a Volume $= \pi r^2 h = \pi \times 4^2 \times 6 = 96\pi$ cm³
b Volume of crate = diameter of 3 cylinders × diameter of
2 cylinders × height
$= 24 \times 16 \times 6 = 2304$ cm³
Volume of 6 cylinders $= 96\pi \times 6 = 576\pi$ cm³
Volume of space = 2304 − 576π cm³
2 Volume of base $= 4 \times 3 \times 1.5 = 18$ cm³
Volume of pyramid $= \frac{1}{3} \times$ area of base × height
$= \frac{1}{3} \times (3 \times 4) \times 6 = 24$ cm³
Total volume = 42 cm³

3 By similar triangles, $\dfrac{h}{8} = \dfrac{5}{4}, h = \dfrac{5}{4} \times 8 = 10$ cm

Volume of large cone $= \frac{1}{3}\pi \times 8^2 \times 10 = 670.20...$
$= 670.2$ cm³ (1 d.p.)
Volume of small cone $= \frac{1}{3}\pi \times 4^2 \times 5 = 83.77...$
$= 83.8$ cm³ (1 d.p.)
Volume of frustum = 670.2 − 83.8 = 586.4 cm³ (1 d.p.)

4 a cyclic

b angle CBX = 180° – (80° + 25°) = 75° (angle sum of
triangle CBX)

angle CBA = 180° – angle CBX (angles on a straight
line)

= 180° – 75° = 105°

angle ADC + angle CBA = 180° (opposite angles in a
cyclic quadrilateral)

$x + 105° = 180°$

$x = 180° – 105° = 75°$

5

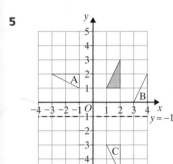

6 a Reflection in the line $y = -1$

b i $\frac{1}{3}$ **ii** (6, 3)

7 a i $\overrightarrow{QP} = \overrightarrow{QO} + \overrightarrow{OP}$

$= -\overrightarrow{OQ} + \overrightarrow{OP}$

$= -(3\mathbf{a} + \mathbf{a}) + 6\mathbf{b}$

$= 6\mathbf{b} - 4\mathbf{a}$

ii $\overrightarrow{RM} = \overrightarrow{RQ} + \overrightarrow{QM}$

$= \overrightarrow{RQ} + \frac{1}{2}\overrightarrow{QP}$

$= \mathbf{a} + \frac{1}{2}(6\mathbf{b} - 4\mathbf{a}) = \mathbf{a} + 3\mathbf{b} - 2\mathbf{a}$

$= 3\mathbf{b} - \mathbf{a}$

b $\overrightarrow{RS} = \overrightarrow{RO} + \overrightarrow{OS}$

$= -\overrightarrow{OR} + \overrightarrow{OS}$

$= -3\mathbf{a} + (6\mathbf{b} + 3\mathbf{b})$

$= 9\mathbf{b} - 3\mathbf{a}$

$= 3(3\mathbf{b} - \mathbf{a}) = 3\overrightarrow{RM}$

Since RS is a multiple of RM, the points R, M and S lie
on a straight line.

8 a 6 km = 6000 m

number of laps = 6000 ÷ 240 = 25 laps

b speed = $\frac{\text{distance}}{\text{time}} = \frac{240}{30} = 8$ m/s

c 6 km = 6000 m

time = $\frac{\text{distance}}{\text{speed}} = \frac{6000}{8} = 750$ s

750 s = 750 ÷ 60 = 12.5 minutes = 12 min 30 s

9 a

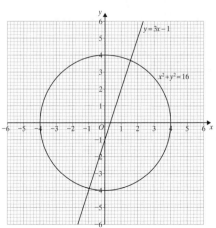

b $x = 1.6, y = 3.7$

and $x = -1, y = -3.9$

(the points of intersection)

10 $AC^2 = AB^2 + BC^2$

$8^2 = 5^2 + BC^2$

$64 = 25 + BC^2$

$39 = BC^2$

$BC = \sqrt{39}$ cm

11 $x^2 = 2.5^2 + 7^2$

$x^2 = 6.25 + 49$

$x^2 = 55.25$

$x = \sqrt{55.25} = 7.43... = 7.4$ m (1 d.p.)

12 a The corresponding sides are in the same ratio,

BC : DE = 2.5 : 15 = 1 : 6

$\frac{AC}{AE} = \frac{1}{6}$

$\frac{3}{AE} = \frac{1}{6}$

$\frac{AE}{3} = \frac{6}{1}$

$3 \times \frac{AE}{3} = \frac{6}{1} \times 3$

$AE = \frac{6}{1} \times 3 = 18$ cm

$CE = AE - AC = 18 - 3 = 15$ cm

b In triangle ABC,

$\tan x = \frac{opp}{adj}$

$\tan x = \frac{2.5}{3} = 0.833333...$

$x = \tan^{-1} 0.833333...$

$x = 39.80... = 39.8°$ (1 d.p.)

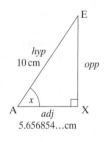

13 Required angle is angle EAC

Using the base of the pyramid:

In triangle ABC,

$AC^2 = AB^2 + BC^2$

$AC^2 = 8^2 + 8^2$

$= 64 + 64 = 128$

$AC = \sqrt{128}$ cm

$AX = \frac{AC}{2} = \frac{\sqrt{128}}{2} = 5.656854...$ cm

In triangle EAX,

$\cos x = \frac{adj}{hyp}$

$\cos x = \frac{5.656854...}{10} = 0.565685...$

$x = \cos^{-1} 0.565685...$

$x = 55.55... = 55.6°$ (1 d.p.)

angle EAC = angle EAX = 55.6° (1 d.p.)

14 Angle AON = 45° + 90° = 135°

Using the cosine rule in triangle AON,

$AN^2 = 4.7^2 + 6.4^2 - 2 \times 4.7 \times 6.4 \times \cos 135°$

$AN^2 = 22.09 + 40.96 - 60.16\cos 135°$

$= 63.05 - (-42.53954396...)$

$= 105.589544...$

$AN = 10.27... = 10.3$ km (3 s.f.)

ANSWERS

Section 1: Algebra

1 Symbols, sequences and equations

1 $3x^2 + 6x^3$
 $= 3x^2(1 + 2x)$

2 **a** $v = -6 + (2.4 \times 1.9)$
  $v = -6 + 4.56$
  $v = -1.44$

 b **i** $5x - 3 = 2 + 3x$
   $2x - 3 = 2$
   $2x = 5$
   $x = 2.5$

  ii $3(2x + 1) = 7 - 5x$
   $6x + 3 = 7 - 5x$
   $11x + 3 = 7$
   $11x = 4$
   $x = \frac{4}{11}$

  iii $\frac{x + 7}{x + 4} = 3$
   $x + 7 = 3(x + 4)$
   $x + 7 = 3x + 12$
   $7 = 2x + 12$
   $-5 = 2x$
   $x = -2.5$

3 **a** $(3 \times 2) + (2 \times -5)$
  $= 6 + (-10)$
  $= -4$

 b $\frac{2 - (2 \times -5)}{3}$
  $= \frac{2 - (-10)}{3}$
  $= \frac{12}{3}$
  $= 4$

4

x	$x^2 - x$	Comment
3	$9 - 3 = 6$	Too low
4	$16 - 4 = 12$	Too high
3.5	$12.25 - 3.5 = 8.75$	Too low
3.6	$12.96 - 3.6 = 9.36$	Too high
3.55	$12.6025 - 3.55 = 9.0525$	Too high

 $x = 3.5$ (1 d.p.)

5 **a** 1, 4, 9, 16, 25 are the square numbers ($1^2 = 1$, $2^2 = 4$ etc.)
  So the nth term is n^2
 b The sequence $-3, 0, 5, 12, 21, \ldots$ is 4 less than the
  sequence in part (a).
  So the nth term is $n^2 - 4$

6 **a** $10a^{2+3} = 10a^5$
 b $\frac{18p^4}{9p^2} = 2p^2$
 c $21g^3h^5$
 d $6a^{-1}b^2$ or $\frac{6b^2}{a}$

7 **a** $4(5m - 3) + 2(6 - m) = 20m - 12 + 12 - 2m = 18m$
 b $n^2 - 49 = (n + 7)(n - 7)$ (Difference of two squares)
 c $w = s^2 - x$
   $w + x = s^2$
   $\sqrt{w + x} = s$
   $s = \sqrt{w + x}$
 d $a(x - b) = a^2 + bx$
   $ax - ab = a^2 + bx$ (removing brackets)
   $ax = a^2 + bx + ab$
   $ax - bx = a^2 + ab$ (separating terms in x from those not
   $x(a - b) = a^2 + ab$ containing x)
   $x = \frac{a^2 + ab}{a - b}$ (taking x as a factor of left-hand side)
      (dividing both sides by $(a - b)$)

2 Coordinates and graphs

1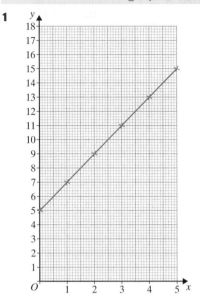

The graph passes through (0, 5) with a gradient of 2

2 If A is the point (0, 4) and B is the point (2, 0),
 a $\left(\frac{0 + 2}{2}, \frac{4 + 0}{2}\right) = (1, 2)$
 b Gradient $= \frac{\text{change in vertical distance}}{\text{change in horizontal distance}}$
  $= \frac{0 - 4}{2 - 0} = -2$
 c Gradient $= -2$, intercept $= 4$
  (passes through (0, 4))
  Equation is $y = -2x + 4$
 d $y = -2x - 1$

3 A = (3, 2, 5), so centre of base X = (3, 2, 0).
 X is midpoint of OB, so B = (6, 4, 0)

4 Gradient $= \frac{8 - 4}{0 - (-2)} = \frac{4}{2} = 2$

 Intercept is at (0, 8)
 Equation is $y = 2x + 8$

3 Quadratic functions

1 **a b**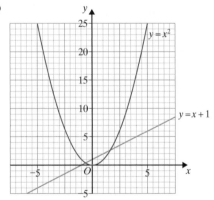

 c $x = -0.6$ or $x = 1.6$
 d $x^2 = x + 1$, or $x^2 - x - 1 = 0$

2 $x^2 - 4x - 2 = 0$
$(x - 2)^2 = x^2 - 4x + 4$
So $(x - 2)^2 - 6 = x^2 - 4x - 2$
$(x - 2)^2 - 6 = 0$
$(x - 2)^2 = 6$
$x - 2 = \pm\sqrt{6}$
$x = 2 \pm \sqrt{6}$

3 a $x^2 - 5x - 24 = (x - 8)(x + 3)$
 b $(x - 8)(x + 3) = 0$
$x - 8 = 0 \text{ or } x + 3 = 0$
$x = 8 \text{ or } x = -3$

4 $x^2 - 4x + 3 = 0$ can be written as $x^2 - 2x + 7 = 2x + 4$ (by adding $2x + 4$ to both sides).
So the required graph is $y = 2x + 4$

5 a $x^2 - 2x - 4 = 0$
$(x - 1)^2 = x^2 - 2x + 1$, so $(x - 1)^2 - 5 = x^2 - 2x - 4$
$(x - 1)^2 - 5 = 0$
$(x - 1)^2 = 5$
$x - 1 = \pm\sqrt{5}$
$x = 3.24 \text{ or } -1.24 \text{ (2 d.p.)}$
 b $3x^2 + 5x - 2 = 0$
$(3x - 1)(x + 2) = 0$
$3x - 1 = 0 \text{ or } x + 2 = 0$
$x = \frac{1}{3} \text{ or } x = -2$
 c $2x^2 - x - 2 = 0$
$a = 2, b = -1, c = -2$
$x = \dfrac{-(-1)\pm\sqrt{(-1)^2 - 4 \times 2 \times -2}}{2 \times 2}$
$x = \dfrac{1 \pm \sqrt{17}}{4}$
$x = 1.28 \text{ or } x = -0.78$

4 Further graphs

1

x	-2	-1	0	1	2	3
y	0.11	0.33	1	3	9	27

(Values given to 2 d.p.)

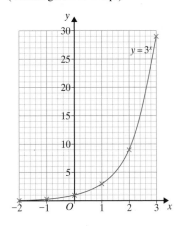

2   Red graph $y = x^2 + 1$
Blue graph $y = (x - 2)^3$
Green graph $y = \frac{1}{x}$

3

x	-2	-1	0	1	2	3	4	5
$x^2 + x$	2	0	0	2	6	12	20	30
2^x	0.25	0.5	1	2	4	8	16	32

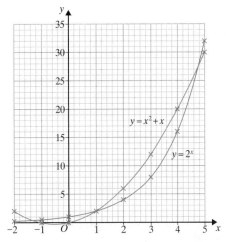

The equation has 3 solutions, as the graphs intersect at 3 points.

5 Simultaneous equations and inequalities

1 $4x + 3 < 1$
$4x < -2$
$x < -0.5$

2 $2x - 3y = 7$ ①
$4x + y = 0$ ②

$2 \times$ ①: $4x - 6y = 14$
②: $4x + y = 0$ (Subtract)
$-7y = 14$
$y = -2$

①: $2x - 3y = 7$
$2x + 6 = 7$ (Substituting $y = -2$)
$2x = 1$
$x = \frac{1}{2}, y = -2$

3

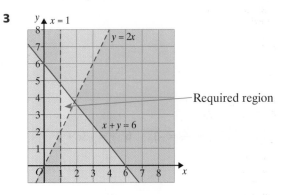

Required region

4
$$x^2 + y = 15 \qquad ①$$
$$x + y = 3 \qquad ②$$
From ②: $\quad x = 3 - y \qquad ③$
$(3 - y)^2 + y = 15 \qquad$ (Substituting ③ into ①)
$9 - 6y + y^2 + y = 15$
$y^2 - 5y - 6 = 0$
$(y - 6)(y + 1) = 0$
$\quad y - 6 = 0$ or $y + 1 = 0$
$\quad y = 6$ or $y = -1$

From ②: $\quad$ If $y = 6, x = -3$
$\qquad\qquad$ If $y = -1, x = 4$

6 Proof

1 a Let the numbers be $n, n + 1$ and $n + 2$
$n + n + 1 + n + 2 = 3n + 3 = 3(n + 1)$, which is a multiple of 3
b Let the numbers be $2n, 2n + 2$ and $2n + 4$
$2n + 2n + 2 + 2n + 4 = 6n + 6 = 6(n + 1)$, which is a multiple of 6

2 No, 2 is a prime number, but the sum of two consecutive numbers is always odd.

3 Let the original number be n.
Then $n \times n = n^2$
But $(n - 1)(n + 1) = n^2 + n - n - 1 = n^2 - 1$
So it works for any number.

Note: by using algebra, we have proved the rule always works, whether n is positive, negative or zero, integer or fraction.
For example $0^2 = 0$, and $-1 \times 1 = -1$, and $2.4^2 = 5.76$, and $1.4 \times 3.4 = 4.76$

Section 4: Algebra

ANSWERS TO EXAMINATION STYLE QUESTIONS

1 $3x(3x - 2)$

2 $\dfrac{3x^2 + 5x - 2}{x^2 - 4} = \dfrac{(3x - 1)(x + 2)}{(x + 2)(x - 2)} = \dfrac{3x - 1}{x - 2}$

3 $s = \dfrac{p}{4} - 2$
$s + 2 = \dfrac{p}{4}$
$4(s + 2) = p$

4 $\dfrac{2x}{3} - \dfrac{x}{2} = 3$
$\dfrac{4x}{6} - \dfrac{3x}{6} = 3$
$\dfrac{x}{6} = 3$
$x = 18$

5 a Gradient $= \dfrac{11 - 2}{-3 - 0} = \dfrac{9}{-3} = -3$ $\quad$ **b** $y = -3x + 2$

$\quad$ **c** $y = \frac{1}{3}x + 2$

6 $(x + 6)(x - 2) = 0$
$x = -6$ or $x = 2$

7 $(x - 3)^2 = x^2 - 6x + 9$
So $(x - 3)^2 - 5 = x^2 - 6x + 4$
$a = 3, b = -5$

8 $a = 1, b = -4, c = -2$
$$x = \frac{-(-4) \pm \sqrt{(-4)^2 - 4 \times 1 \times -2}}{2 \times 1} = \frac{4 \pm \sqrt{16 + 8}}{2} = \frac{4 \pm \sqrt{24}}{2}$$
$= 4.45$ or -0.45 (2 d.p.)

9 $\dfrac{5}{x - 2} + \dfrac{x}{x + 2} = 2$ $\quad$ Multiply each term by $(x - 2)(x + 2)$

$$\frac{5(x - 2)(x + 2)}{x - 2} + \frac{x(x - 2)(x + 2)}{x + 2} = 2(x - 2)(x + 2)$$
$$5(x + 2) + x(x - 2) = 2(x - 2)(x + 2)$$
$$5x + 10 + x^2 - 2x = 2x^2 - 8$$
$$x^2 + 3x + 10 = 2x^2 - 8$$
$$0 = x^2 - 3x - 18$$
$$(x - 6)(x + 3) = 0$$
$$x = 6 \text{ or } x = -3$$

10 a

x	-3	-2	-1	0	1	2	3
x^3	-27	-8	-1	0	1	8	27
$-x$	3	2	1	0	-1	-2	-3
y	-24	-6	0	0	0	6	24

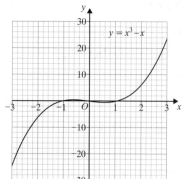

b $x = 1.5$

11

x	-3	-2	-1	-0.5	0	0.5	1	2	3
y	0.11	0.25	1	4	$-$	4	1	0.25	0.11

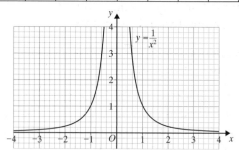

12
$$3x + 2y = 5 \qquad ①$$
$$4x + 3y = 6 \qquad ②$$

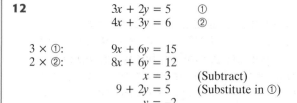

$3 \times ①$: $\qquad 9x + 6y = 15$
$2 \times ②$: $\qquad 8x + 6y = 12$
$\qquad\qquad\qquad x = 3 \qquad$ (Subtract)
$\qquad\qquad 9 + 2y = 5 \qquad$ (Substitute in ①)
$\qquad\qquad\qquad y = -2$

13 Let the even number be $2n$.
Then the even number squared: $(2n)^2 = 4n^2$
Which is a multiple of 4